Philipp Rüger

Automated assessment in viticulture

Philipp Rüger

Automated assessment in viticulture

Construction of a robotic vehicle for high-throughput phenotyping of grapevines

ScienciaScripts

Imprint

Cover image: www.ingimage.com

This book is a translation from the original published under ISBN 978-3-8381-5306-3.

Publisher:
Sciencia Scripts
is a trademark of
Dodo Books Indian Ocean Ltd. and OmniScriptum S.R.L publishing group

120 High Road, East Finchley, London, N2 9ED, United Kingdom
Str. Armeneasca 28/1, office 1, Chisinau MD-2012, Republic of Moldova, Europe
Managing Directors: Ieva Konstantinova, Victoria Ursu
info@omniscriptum.com

Printed at: see last page
ISBN: 978-620-8-98851-7

Table of contents

1 Introduction

The history of viticulture is characterized by constant change. Over the course of time, changes have taken place more or less rapidly. Naturally, these are closely linked to the development of society. As the pace of human development accelerates, so do the changes in areas of viticulture. If we look at antiquity and the Middle Ages, the innovations in these eras were rather moderate. With the beginning of industrialization, the general development virtually exploded, with all the positive and negative effects, as we know today. Technical progress did not stop at viticulture either. Cultivation and management of the vineyards, as well as the ageing of the wine, have been enormously simplified and made easier. This can be measured, among other things, by the reduction in the number of man-hours required per unit of land. While these were in the range of over 1,000 man-hours (AKh) in the days of purely manual cultivation, today a vineyard can be cultivated with less than 100 AKh per hectare (ha), depending on the degree of mechanization (KTBL 2013). In addition to all the positive aspects, experience has shown that there are also negative consequences. The availability of cheap and safe means of transport made it possible for a broad mass of people to travel, especially from the 19th century onwards. People were also happy to cross continental borders. Naturally, agricultural goods and plants were also exchanged. In the course of thousands of years of cultivation, plants in particular have adapted to their surrounding ecosystem and developed natural defense mechanisms against pests. In North America, fungi such as *Plasmopara viticola* or *Erysiphe necator*, colloquially known as downy and powdery mildew, are native. The vines growing there have developed resistance mechanisms against these fungi. In the middle of the 19th century, these pathogens were introduced to Europe, where the vines are not resistant to them. The situation is similar with phylloxera, although grafting technology has provided a remedy here. In the area of fungal pathogens, there is still a need for high fungicide expenditure to combat the diseases in Europe's native grape varieties. By selectively crossing the resistance mechanisms of American vines into European varieties, it is possible to breed so-called fungus-resistant varieties.

1.1 Necessity

The time frame required for breeding a new grape variety is usually between 25 and 30 years (KADISCH et al. 2008). The reasons for this long period of time are, on the one hand, the generative and vegetative multiplication of the seedlings in question and, on the other hand, the time-consuming bonitures during this time. On the one hand, the desired resistance to the relevant pathogens and, on the other hand, viticultural parameters such as yield level, growth behavior and sensory properties are checked. Suitable scoring schemes are used, for example according to OIV (EUROPEAN UNION 2012) or BBCH (BBA 2001). This is usually done manually, i.e. professionally trained inspectors are used. Their deployment incurs high costs, among other things. They can only be deployed for a relatively short time (2-4 hours) due to the high level of concentration required, and the scoring results are also heavily influenced by subjective factors. As a result, these methods are often only used to test selected breeding lines and greenhouse cultures, although it would be desirable to use them across a wide area in order to shorten the long period between the start of breeding and the approval of the grape variety through better scoring (ROSCHER et al. 2014).

For these reasons, the "PHENOvines" research project called for by the Federal Ministry of Education and Research is dedicated to the development of a process for "high-throughput phanotyping of yield parameters and mildew infestation in grapevines". The aim is to automate the scoring process, starting with the recording of phenotypic characteristics in the culture, through further processing to the analysis of the data. Thematically, the project is divided into the sub-projects A "Phanotyping in vine stocks" and B "Construction of a robot". Project part A is under the responsibility of the Julius Kühn Institute (JKI), Institute for Grapevine Breeding, Geilweilerhof site, and deals with the core issues of "image acquisition" and "image analysis". The Institute of Technology at Geisenheim University is working on project part B at and is thus responsible for the development of the automatically moving carrier platform that carries the image-capturing camera unit. This part is also the core of this book. The focus here is on the conception and design of the robot platform; experimental data collection only serves to validate the design.

1.2 Theses

As part of the project application, target definitions were drawn up according to which the purpose and characteristics of the scoring system to be created were clearly defined. The requirements profile was based on the conditions at the Institute for Grapevine Breeding, Geilweilerhof, as it was planned to incorporate the scoring platform into the work program at the site during the course of the project. According to these specifications, key data and load profiles were defined to which the development is aligned.

Based on these planning principles, the following theses were formulated, taking into account general viticultural requirements and the current state of knowledge in the field of agricultural robotics:

I. Taking into account general viticultural requirements, the PHENOvines project is developing an automatically moving platform for use in vineyards.

II. The PHENObot scoring platform can be used to navigate vineyard alleys with sufficient accuracy, ensuring safe and reproducible navigation within defined operating limits.

III. In order to carry out the scoring task, existing stock coordina- tions are automatically read into a data processing system and processed in such a way that driving commands and work orders are generated from them. The necessary software applications are part of the scoring system and are specially adapted to it.

IV. The scoring system is based on an automatically moving carrier vehicle. This is equipped with an adaptive storage control system with four degrees of freedom to accommodate image-capturing sensors.

V. The scoring system enables the powerful automated phanoty- pization of vine stocks with an area output of at least 250 vines per hour.

The examination and validation of these five theses is a central part of this thesis.

2 Literature

In the following, the basics of positioning systems, especially satellite-based systems, and geodata are discussed. Possible applications in technology, agriculture and viticulture are shown. Further chapters deal with the developments and applications in robotics and phanotyping and remote sensing applications.

2.1 Positioning systems and their application

Satellite-based positioning systems are used in many areas of life today and have become almost indispensable. Among the many possible applications are

- the military and security sector,
- the transportation (land, water and air vehicles),
- Fleet management,
- the construction and surveying,
- the leisure sector,
- Science and research,
- agricultural applications.

Each of these areas has its own requirements and requires different levels of precision. The basics of surveying and satellite-based positioning are explained below. Section 2.1.3 shows the structure and possible applications of geoinformation systems. Section 2.2 deals specifically with applications in the agricultural sector.

2.1.1 Geodetic principles

MANSFELD (2010) fundamentally divides *positioning* into two groups: *positioning* and *navigation* (Table 1). Satellite positioning was first used for navigation in the military sector and was only used for positioning after it was approved for civilian use (MANSFELD 2010).

Table 1 Applications of localization, modified according to MANSFELD (2010).

Positioning	***Navigation***
Geodesy	Space travel
Object measurement	Aviation

BAUER (2011) places the terms *surveying, localization* and *geodesy* on the same level and sees the distinguishing features of the first two in the point dynamics. The task of *surveying* is to describe the geometry of objects using suitable methods, whereby the points to be recorded are in a relative rest position to the earth's surface. In contrast, locating generally involves recording points that are in motion relative to the earth's surface or that can be in motion. This means that positioning can be seen as part of *navigation* (BAUER 2011). BAUER (2011) also quotes the German geodata EXPERT HELMERT: "Geodăsie is the science of measuring and mapping the earth's surface". However, this must take into account all the tasks that geodata have to cope with today, including the observation of geodynamic phenomena and the determination of the Earth's gravitational field (BECKER & HEHL 2012). BAUER (2011) summarizes the difference between *geodesy* and *surveying* as follows: "Geodesy provides the theoretical basis for surveying practice".

The classic method of surveying is *triangulation*. On the one hand, the so-called *fixed point fields* can be generated, and on the other hand, coordinates of new points or distances between points can be determined with the help of existing fixed point fields (MOSER et al. 2012). With this method, the earth can be assumed to be a flat system in the short-range area, while the *figure of the earth*, i.e. the curvature of the earth, must be taken into account for long-range measurement tasks. Before the availability of satellite geodesy, there were no globally uniform geodetic point fields, as the coasts of the oceans represented a natural, insurmountable limit for the *triangulation* method. For example, the distance between Europe and America could only be determined to an accuracy of around 100 m (BAUER 2011). With the advent of the satellite age in the 1950s and 1960s, these limits were overcome. BECKER and HEHL (2012) see three cornerstones of geodesy:

- Geometry,
- Gravity field and geoid,
- Earth's rotation and orientation.

These are linked together by a reference system (BECKER & HEHL 2012). With a *reference system* it is basically possible to determine the spatial position, the location, the height or the gravity of points. It is generally divided into

- Position reference systems,
- High reference systems,
- Heavy-duty cover systems

are distinguished. All consist of a coordinate system and a fixed point field. *Reference planes* are derived from the respective *reference systems.* For measurements with geodetic instruments, these are always aligned with their vertical axis in the direction of the center of gravity. The resulting planes perpendicular to the gravity vector are referred to as *aquipotential planes* or *level planes.* In geodesy, these are related to the mean sea level and are referred to as *geoids* (Fig. 1). Due to mass irregularities in the earth's interior, there are *perpendicular deviations* when aligning the measuring rates in comparison to the earth's surface; in relation to the idealized representation as an ellipsoid, the *perpendicular* results in contrast to the *vertical.* This results in the figure in Fig. 1, which has bumps and dents. Mathematically, the *geoid* cannot be described as a function. However, compared to the earth's size, the *geoid* approximates the idealized shape of the rotational ellipsoid well. The deviation between the two is referred to as *geoid undulation* or *geoid height* (Fig. 2) (BAUER 2011, MOSER et al. 2012, BECKER & HEHL 2012). Depending on the shape of the geoid, resulting from the mass distribution in the Earth's body, the *geoid undulation* assumes positive or negative values. In the Geisenheim area, the undulation is 48.34 m (BRUNNER 2014), in the Indian Ocean, for example, it is -100 m (in the WGS84 reference system) (BAUER 2011).

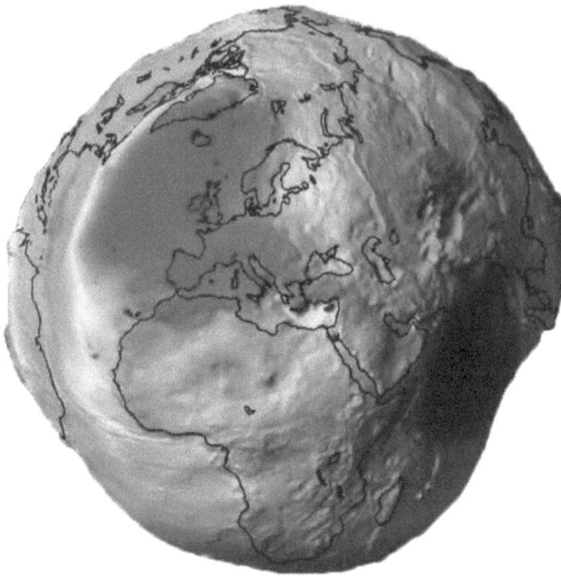

Fig. 1 The geoid modeled by the Gravity field and steady-state Ocean Circulation Explorer (GOCE). The areas of positive geoid undulation are shown in red, the negative areas in blue (ESA 2013).

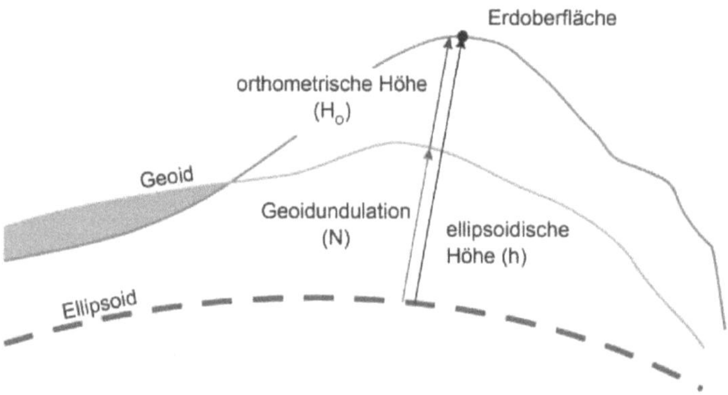

Fig. 2 Relationship between ellipsoidal height, geoid undulation and orthometric height, modified after BKG (2014b).

The rotational ellipsoid used is referred to as the *reference ellipsoid* and has its center of gravity at the center of mass of the Earth. It is defined by the parameters

- Large half-axis (a),
- Small half-axis (b),
- Flattening *(f)*

described. These are in the relationship

$$f = \frac{a-b}{a} \qquad \text{Gl. 2-1}$$

There are a number of historical and important *reference ellipsoids* (Table 2) (MOSER et al. 2012):

Table 2 Parameters of some important reference ellipsoids (MOSER et al. 2012).

Reference ellipsoid	***Large half-axis a [m]***	***Small half-axis b [m]***	***Flattening***
BESSEL	6.377.397	6.356.078	1:299,1528
HAYFORD	6.378.388	6.356.911	1:297,0000
KRASSOVSKY	6.378.245	6.356.863	1:298,3000
WGS72	6.378.135	6.356.750	1:298,2600
GRS80	6.378.137	6.356.752	1:298,2572
WGS84	6.378.137	6.356.752	1:298,2572

Points and/or higher geometric information are described spatially as coordinates in the form of number pairs or number triples. For geodetic use, this information must be linked to the earth body (BAUER 2011). *Reference systems are used* for this purpose. The basis of such a reference frame is a *reference ellipsoid*. In this, *coordinates* are represented in a specific system and in relation to a *reference point* (BAUER 2011, BECKER & HEHL 2012, MOSER et al. 2012).

Two systems are generally used as coordinate types. The right-angled, three-dimensional *Cartesian coordinates* use orthogonal axes of the same scale with the designations *X, Y* and *Z* (Fig. 3). In a modification, this system can also be used as a *topocentric coordinate system* for better comprehensibility. In this case, it is applied tangentially to the surface of the earth or the *geoid* or the *rotation elipsoid*. The vertical axis *Z* then coincides with the earth's radius or the *ellipsoid normal*. The orientation of the other two axes points north and east with the designations *Northing* and *Easting* respectively. The choice of the primary axis differs in the fields of *navigation* and *geodesy*. In the former, the first axis usually points to the east and the second to the north. The *Geodăsie* works with reversed axes, with the primary axis pointing north and the secondary axis pointing east. In navigation applications, angles are displayed counterclockwise, while in *geodetic navigation they* are displayed clockwise. Thus, *navigation* corresponds in principle to the usual mathematical conventions, while *geodesy* uses the opposite orientation (BECKER & HEHL 2012).

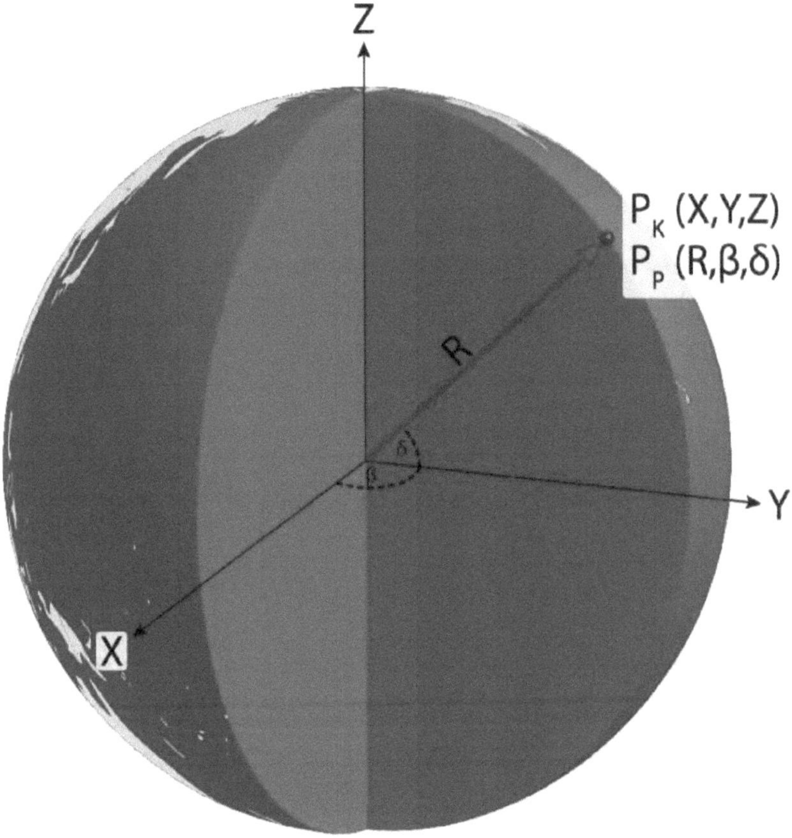

Fig. 3 Three-dimensional Cartesian coordinates (P_k) and polar coordinates (P_p), modified after MÖSER et al. (2012).

Another frequently used form of representation is *spherical* or *ellipsoidal coordinates*. The coordinates of a point are given as *latitude, longitude* and *altitude*. The *latitude β* is defined as the angle of the *ellipsoid normal R to the XY plane*, the *longitude δ* as the angle of the meridian plane through the point *P* to the *XZ plane* through the *prime meridian* at Greenwich (BECKER & HEHL 2012).

For *Cartesian coordinates*, meters are used as the unit; for *ellipsoidal coordinates*, sexagesimal degrees, minutes, seconds or decimal degrees are used for the angular units. In the case of *ellipsoidal coordinates*, the height is also given in meters.

Geodetic coordinates are projected into the plane for visualization and easier processing. This flattening results in a distorted representation of the curved surfaces. Two conformal (true-angle) *projection types* have become established in the German-speaking world. *The Gauß-Kmger projection (GK)* as a *transverse cylinder projection* maps the *contact meridian* true to length and uses 3° wide projection lines (Fig. 4). In the past, it was used as the official coordinate system in Germany (MΘSER et al. 2012, ADV 2015).

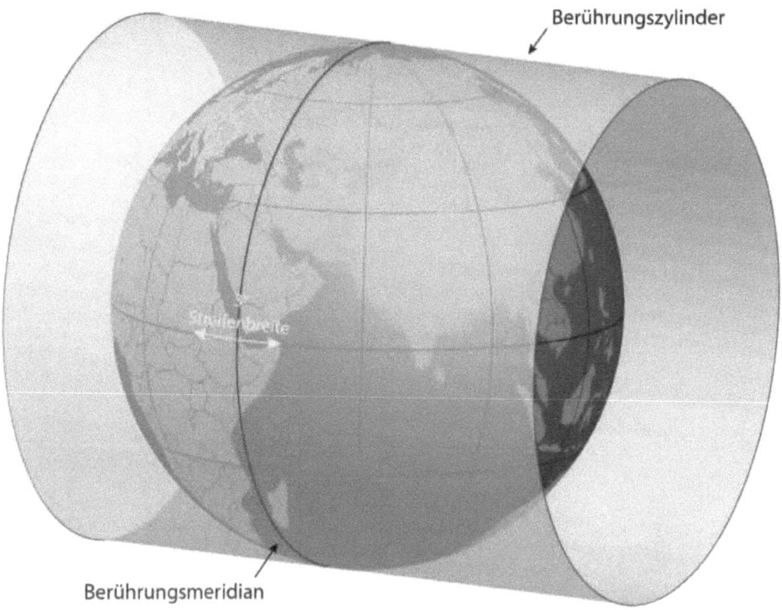

Fig. 4 Schematic representation of the contact cylinder of the Gauß-Krnger mapping, modified according to BVV (2009).

With the availability of a uniform European *reference system*, it was decided to use the *Universal Transverse Mercator projection (UTM)* as the corresponding map (Fig. 5) (ADV 1995). In contrast to GK, the meridian strips of are 6° wide, which means that fewer zones are required for a large east-west extent (Fig. 6) (ADV 2014). The territory of the Federal Republic of Germany lies predominantly in zones 32U and 33U (BECKER & HEHL 2012).

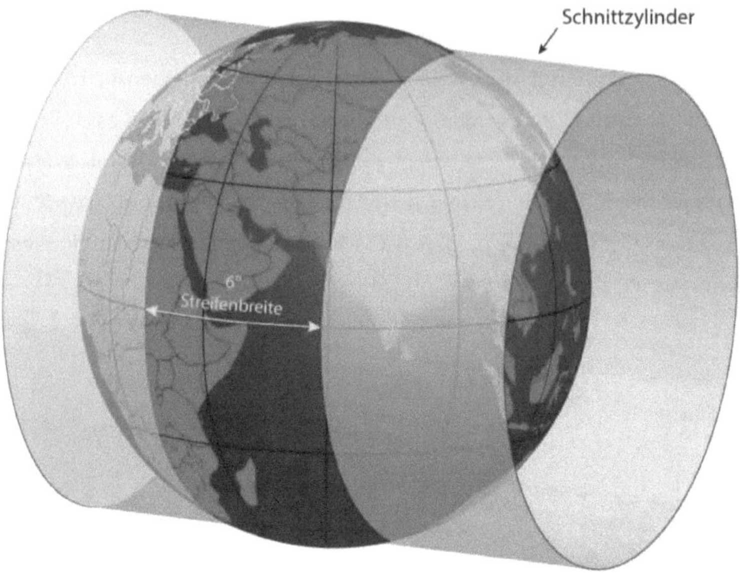

Fig. 5 Schematic representation of the sectional cylinder of the UTM mapping, modified according to BVV (2009).

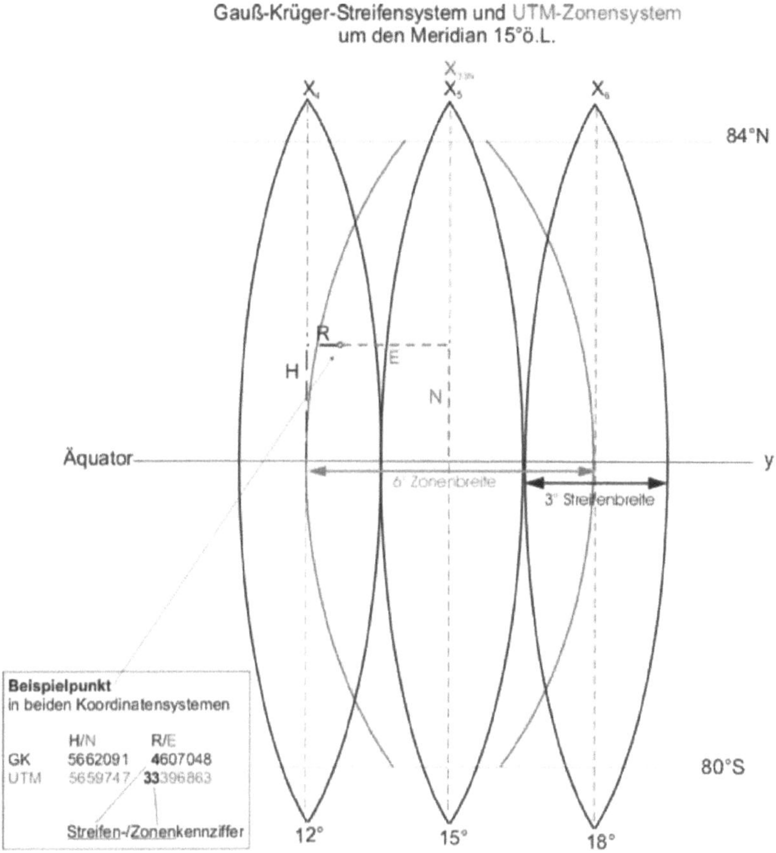

Fig. 6 Gauß-K^ger strip system and UTM zone system (GEOSN 2014).

The coordinate system or the specification of the projection alone does not represent a position reference system. An associated *reference system* or *coordinate reference system (CRS)* must always be specified (Fig. 7) (DIN 2007). An example of this is the *World Geodectic System 1984 (WGS84)* as a global terrestrial reference system. In addition to the mathematical part already described, the coordinate systems, there is also a *geodetic reference system* (physical part), consisting of the above-mentioned *reference point* and a *reference frame* (fixed point fields and velocities). The position of the *reference ellipsoid* in the earth body/geoid, taking into account the orientation and the measuring rod, is referred to as the *geodetic datum* (MΘSER et al. 2012). In summary, it can also be understood as the position of an *eccentric coordinate system* relative to a *geocentric coordinate system*

(BECKER & HEHL 2012).

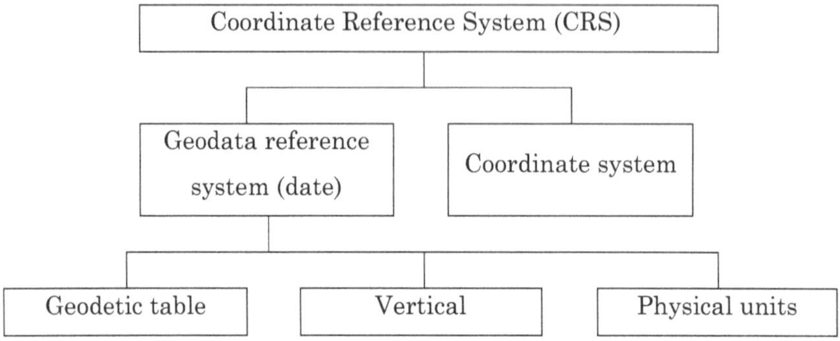

Fig. 7 Definition of the coordinate reference system according to ISO 19111:2007 (DIN 2007).

The *geodetic datum* is described using seven defined parameters:

- 3 translations (ΔX, ΔY, ΔZ),
- 3 rotations (ε_X, ε_Y, ε_Z),
- 1 scale factor (m, q or μ).

In addition, the parameters of the reference ellipse described above apply. Three important transformations are worth mentioning in connection with the *geodetic datum*. *Datum transformation* is the transformation between eccentric and geocentric coordinate systems, *datum transition* is the transformation between two eccentric coordinate systems of different dates and *ellipsoid transition* is the transformation between reference systems of different ellipsoid parameters (BECKER & HEHL 2012, MOSER et al. 2012). The calculations are based on equations that can be found in the relevant collections of formulas (BECKER & HEHL 2012, GRUBER & JOECKEL 2014).

Significant examples of *geodetic datum definitions* are (BAUER 2011, BECKER & HEHL 2012, MÖSER et al. 2012, GEOSN 2014):

- International Terrestrial Reference System (ITRS),
- European Terrestrial Reference System 1989 (ETRS89),
- World Geodectic System 1984 (WGS84),
- Rauenberg Date RD/83.

The ITRS and ETRS89 are based on the reference ellipsoid of the *Geodectic Reference System 80 (GRS80)*, whereby the epoch 1989.0 of the associated reference frame was assumed for ETRS89. WGS84 is basically identical except for a negligible deviation in the flattening (ADAM et al. 2000). Since then, due to plate tectonics, the reference systems in Central Europe have drifted apart by approx. 2.5 cm in a north-easterly direction every year (Fig. 8). The *reference frame* includes the so-called *realizations* of the reference systems, for example the *International Terrestrial Reference Frame 2008 (ITRF2008)*. In these frames, the transformation parameters are regularly recalculated using geodetic spatial methods (BECKER & HEHL 2012, MÖSER et al. 2012).

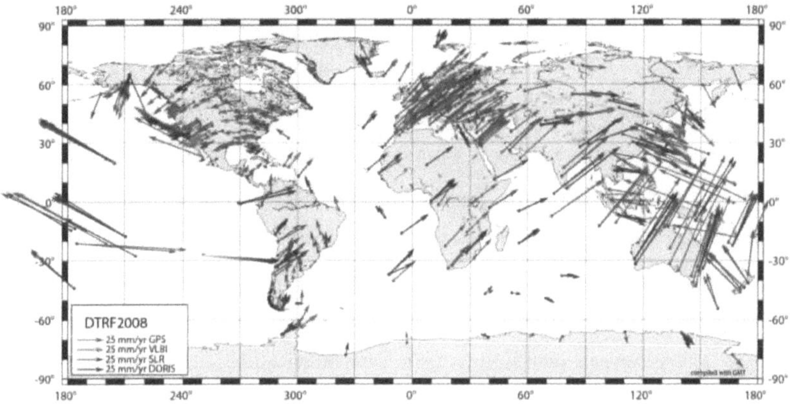

Fig. 8 Velocity field DTRF2008; further realization of the ITRS. The arrows show the annual displacement of the measuring stations as a vector, the length is a measure of the speed (SEITZ et al. 2012).

In practical geodesy, it would be disadvantageous if the coordinate values of the fixed point fields were constantly changing due to plate drift. For this reason, at the beginning of the 1990s, the EUREF (European Reference Frame) working group decided to establish a regional reference system ETRS89 for Europe. The global reference system ITRS of the epoch 1989.0 was chosen as the basis. This eliminates the European continental drift in the ETRS89 reference system (GORRES et al. 2012). Building on this, the Working Group of the Surveying Administrations of the Länder of the Federal Republic of Germany (AdV) decided to introduce ETRS89 with UTM as the associated coordinate system in the area of land surveying and real estate cadastre throughout Germany (ADV 1995). The introduction at state level has largely been completed.

2.1.2 Satellite positioning and navigation

The availability of satellite technology has made it possible to simplify and solve a number of geodata issues. For example, geoinformation can now be specified in global position reference systems.

This decisively increased the accuracy of geodesy and geodynamics (BAUER 2011). The military in particular has placed enormous demands on technical development in this area in order to provide the armed forces with globally unambiguous and universally available positioning and navigation solutions (DODEL & HÄUPLER 2010). Satellite-based positioning systems are generally referred to as *Global Navigation Satellite Systems (GNSS)*. Internationally, the American *Navigation Satellite Timing and Ranging Global Positioning System (NAVSTAR GPS)* and the Russian *Globaluaya Navigatsionnaya Sputnikovaya Sistema, (GLONASS)* are of particular importance. The more commonly used term *NAVSTAR GPS*, or *GPS* for short, has become established as a synonym for satellite-based positioning and navigation applications. However, the acronym *GNSS* is technically correct. There is also the European system *Galileo*, which is currently under development, the Chinese *Beidou*, the Japanese *QZSS* and the Indian *IRNSS*, which is also under development (MANSFELD 2010). All important modern GNSS work as passive systems, i.e. a *receiver* receives signals from several satellites and can calculate its position and speed from them (BAUER 2011). BAUER (2011) summarizes the functions of a GNSS by answering three questions:

- Where am I?
- How fast do I move?
- How late is it?

The positioning works in the same way as the known earth-based radio positioning systems (MANSFELD 2010). A receiver calculates the transit time of the signal transmitted by the transmitter. Trilateration is carried out to determine the position of the receiver unit (Fig. 9). GNSS requires at least four available satellites (SV), with the fourth channel/SV serving to synchronize or correct the time information (KRÜGER et al. 1994).

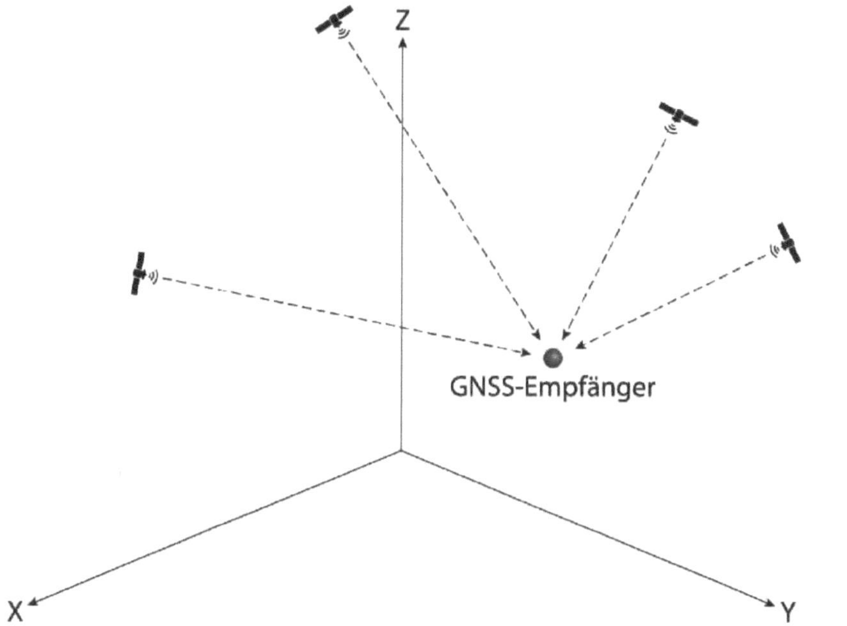

Fig. 9 Basic principle of modern GNSS positioning with at least 4 SV in the global coordinate reference system WGS84 (GPS), modified after B*AUER (2011).*

The position of the receiver is determined by the intersection of the spherical surfaces of the radio signal being emitted (MANSFELD 2010).

The calculation is a system of equations of spherical trigonometry with four equations and four unknowns (DODEL & HAUPLER 2010):

$$d_1 = \sqrt{(x_1 - x)^2 + (y_1 - y)^2 + (z_1 - z)^2} = c \cdot (t_1 - t) \qquad \text{Gl. 2-2}$$

$$d_2 = \sqrt{(x_2 - x)^2 + (y_2 - y)^2 + (z_2 - z)^2} = c \cdot (t_2 - t) \qquad \text{Gl. 2-3}$$

$$d_3 = \sqrt{(x_3 - x)^2 + (y_3 - y)^2 + (z_3 - z)^2} = c \cdot (t_3 - t) \qquad \text{Gl. 2-4}$$

$$d_4 = \sqrt{(x_4 - x)^2 + (y_4 - y)^2 + (z_4 - z)^2} = c \cdot (t_4 - t) \qquad \text{Gl. 2-5}$$

The more orthogonal the spherical sections are, the better the achievable accuracy of the localization. The SV spans an inverted pyramid with the receiver at the apex. Their volume is proportional to the accuracy: the smaller the volume, the worse the accuracy. So if all SVs, regardless of their number, are in a line above the receiver, this volume will be zero (DODEL & HAUPLER 2010).

All systems basically work according to this functional principle. For this reason, only NAVSTAR GPS, which has enjoyed the greatest popularity to date, will be discussed below. The technical characteristics also only refer to GPS.

GNSS consist of three segments. In the case of GPS, the *space segment* nominally contains 24 SVs, whereby 30 SVs are currently (December 2014) in operation and one SV is held in reserve (NGA 2014). These are located in six orbits (Fig. 10) with an orbital period of eleven hours and 58 minutes, an inclination of 55° to the aquatorial plane and an orbital distance of 60° (LECHNER & BAUMANN 2000). The orbital altitude is 20,230 km (MANSFELD 2010); at least four SVs are visible to every user on Earth at the same time (NCO 2015c). With the increased number of satellites, the real availability is between five and eight SV (LECHNER & BAUMANN 2000). The identical satellite constellation is visible every day with a delay of approx. four minutes (MANSFELD 2010). The SV transmit the positioning signals as high-frequency carriers in the L-band as an optimum between the necessary bandwidth, radiation characteristics and vaporization. Two transmission frequencies are currently used, the L1 band at 1,575.42 MHz and the L2 band at 1,227.60 MHz.

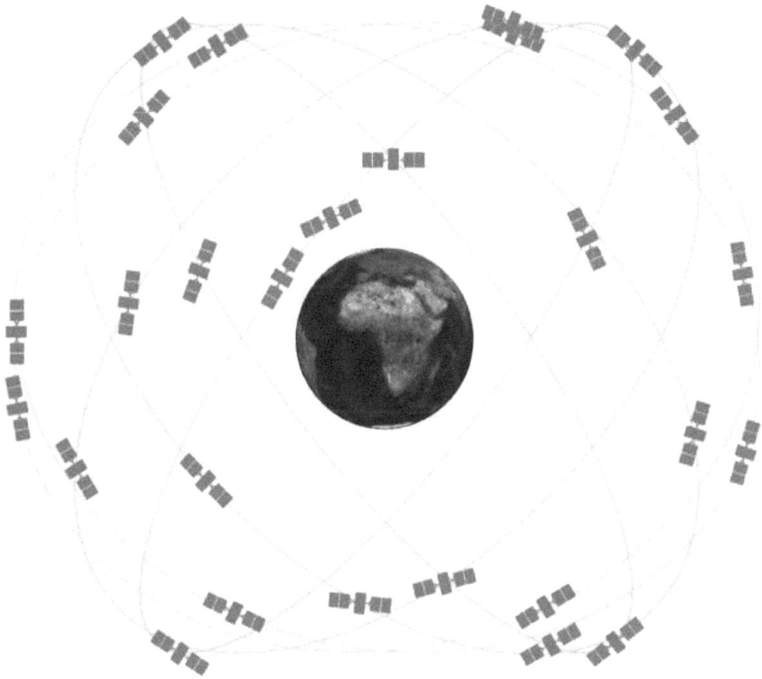

Fig. 10 Orbits of the GPS satellites (NCO 2015c).

These are multiplications of the fundamental system frequency of 10.23 MHz, although this is 4.550 mHz lower due to the elimination of relativistic effects (MANSFELD 2010). The Ll band is coded with the *Coarse Acquisition Code (C/A)* and is publicly available. The L2 band is encoded with the *Precision Code (P(Y)),* which is only accessible to military users (US armed forces). The civilian service is referred to as *Standard Positioning Service (SPS),* the protected service as *Precise Positioning Service (PPS)* (LECHNER & BAUMANN 2000). By recording and evaluating the phase positions of the high-frequency carrier, modern receivers are also able to use the L2 band without knowledge of the P(Y) code and eliminate *ionospheric errors* and *multipath effects* (DODEL & HAUPLER 2010). With the current satellite generations, two further civil signals will become available, the L2c band and the L5 band, each with freely available coding. The upcoming third generation of satellites will then provide a fourth civil signal, the L1c band (NCO 2015c).

DODEL et al. (2010) and MANSFELD (2010) summarize the most important sources of error in the use of GNSS:

- Ionospheric defects,
- Troposhare error,
- Multi-path errors (multipath effects) due to reflections,
- Receiver error,
- Geometry-related errors,
- Incorrect time information from the SV,
- Incorrect path parameters.

These errors have an impact on accuracy in the meter range. The deliberate reduction in measurement accuracy was the so-called *Selective Availability (S/A)* by the US government. The reason for this was the required civilian availability of GPS at the end of the 1980s. In order to nevertheless limit the accuracy in the then freely available civilian range to around 100 m for safety reasons, the SVs from generation IIA onwards were equipped with the ability to simulate stochastically occurring orbit and clock errors. Following a directive from US President Clinton, this function was switched off again on May 2, 2000 (Fig. 11). From Generation III onwards, the GPS satellites will no longer have *S/A* (NCO 2015b).

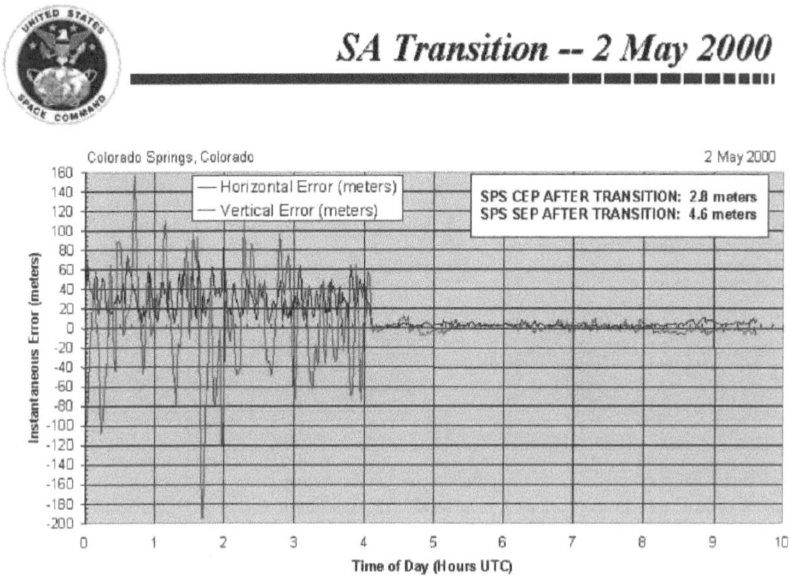

Fig. 11 Shutdown of Selective Availability on May 2, 2000. The accuracy was improved from 100 m to <10 m (NCO 2015b).

The *ground segment* of the GPS fulfills four main tasks (MANSFELD 2010)

- Function check of the entire system,
- Observation of satellite orbits,
- Control of the time information,
- Prediction of satellite ephemeris and time information.

The organ, also known as the control segment (Fig. 12), consists of (NCO 2015a)

- 1 master control station,
- 16 monitor stations (Monitor Station),
- Ground transmitting stations (ground antennas).

Fig. 12 Ground control segment of the NAVSTAR GPS, modified according to NCO (2015a).

The *user segment* consists of all users of a GNSS. All information supplied by the system can be

- Position,
- Speed,
- Time,

with different accuracy criteria, depending on the requirements and technical equipment. Possible areas of application are listed below as examples (no claim to completeness):

- Geodesy/surveying,
- Transportation (ground, air, sea),
- Military and security sector,
- Space travel,
- Science,
- Agriculture (section 2.2).

Positioning and navigation systems can be described using *performance parameters*. There is an abundance of these as well as other influencing variables. For reasons of clarity, only three important terms will be discussed here:

- Visibility,
- Availability,
- Integrity.

Visibility initially only indicates whether, for example, an SV is visible at the user's location, i.e. that a line of sight between the two is free of obstacles. However, connecting lines close to the horizon must be evaluated critically, as topography can often cause obstructions. In addition, with this geometry the positioning signals have to pass through the atmosphere over a longer distance, which can result in propagation time errors (MANSFELD 2010). For this reason, a minimum *elevation* angle is defined. Below this limit, visible SVs are ignored by the receiver. In practice, the *elevation mask* is usually between 5° and 15° (KRÜGER et al. 1994, LECHNER & BAUMANN 2000, DODEL & HAUPLER 2010). To check visibility, software applications can be used that can indicate the visibility of SVs depending on time, location and elevation. An example of this is the online application from Trimble: http://www.trimble.com/GNSSPlanningOnline/

Availability indicates that at least four SVs are available to the user, which in principle enables positioning using GNSS. Combined with a required accuracy, *availability* can be characterized with the *Dilution of Precision (DOP)* factors. In general, the DOP factors are a measure of the geometrically induced positioning errors. The following applies

$$DOP = \frac{Standardabweichung\ des\ Positionsfehlers\ \sigma_p}{Standardabweichung\ des\ Entfernungsfehlers\ \sigma_r} \quad \text{Gl. 2-6}$$

Further differentiations of the DOP factors are made (MANSFELD 2010):

- PDOP(Position DOP),
- HDOP(Horizontal DOP),
- VDOP(Vertical DOP),
- TDOP(time DOP).

In relation to the volume of the above pyramid, this means that The larger the volume (up to the maximum), the higher the accuracy, the smaller the DOP value

(DODEL & HAUPLER 2010). The authors classify the values in Table 3.

Tab. 3 Classification of the DOP values (DODEL & HA UPLER 2010).

DOP	***Grade***	***Service description***
1	Very good	Satisfies the highest demands
2-3	Good	Sufficient for all accuracy requirements
4-6	Satisfactory	Good for general requirements
7-8	Sufficient	Value is acceptable to a limited extent for many applications
9-20	Poor	Value is only suitable as a rough estimate
>20	Insufficient	Measurements are accurate to the size of a soccer field; should not be used

HDOP and PDOP are usually used in GNSS measurement systems as quality parameters for assessing a satellite constellation. TRIMBLE (2001), for example, recommends a PDOP of six as the upper bound for a *real-time kinematic measurement* (TRIMBLE NAVIGATION LIMITED 2001).

The *integrity* of a system characterizes its ability to warn the user of a failure in good time. The factors system accuracy at the tolerance limits, time to an alarm, integrity risk and false alarm rate are included (DODEL & HAUPLER 2010, MANSFELD 2010).

The achievable accuracy when using GNSS alone (stand-alone) is not sufficient for a number of applications. For this reason, systems have been established in the past to increase precision and reliability, so-called *augmentation* (DODEL & HÄUPLER 2010). The basics of two methods will be discussed in this overview (KRÜGER et al. 1994, LECHNER & BAUMANN 2000, DODEL & HÄUPLER 2010, MANSFELD 2010).

With the help of *differential measurement methods (DGNSS),* the accuracy of the runtime/distance measurement and thus the position measurement can be significantly improved. In this method, two receivers are used, one for the *user segment (rover)* and the second as a *reference station (base).* The latter is positioned at a location whose coordinates are known from a precise measurement. The *base* is used to measure a position; the known location can be used to eliminate system and propagation-related errors using suitable algorithms. The corrected position can be communicated to the *rover* via a radio signal, the so-called *correction signal.* This reduces the influence of errors. When determining the position, the distance to the SV is measured using two methods: *code phase measurement* and *carrier phase measurement.* The former produces unambiguous results with an accuracy in the meter range, the latter produces ambiguous results with accuracies in the centimeter range. The *code phase measurement* is generally referred to as *DGNSS* or *DGPS*. The correction data is transmitted as a radio signal in the RTCM SC-104 (Radio Technical Committee for Maritime Applications) data format via the GSM, VHF network or the 2 m band (160 MHz). DGNSS is used for augmentation services such as the *Wide Area Augmentation System (WAAS)* and the *European Geostationary Navigation Overlay Service (EGNOS)*. These were developed to ensure sufficient accuracy and security.

(integrity testing) has been developed for civil aviation. Additional techniques for integrity checking are used here. The correction signals are transmitted via satellite technology.

Trager phase measurement uses the *real-time kinematic (RTK)* method with high precision. The reference stations are usually operated directly by the user. These can be mobile or permanently installed facilities. The standard data format of the RTCM (RTCM 2.x or RTCM 3.x) is also used, although individual manufacturers also offer proprietary formats for their systems, e.g. Trimble with CMR and CMR+ (Compact Measurement Record) (TRIMBLE NAVIGATION LIMITED 2001). Public systems are also offered in the field of RTK, such as the *Satellite Positioning Service (SAPOS)* initiated by the AdV. Three services are provided with different accuracies, comparable with user-owned RTK stations up to 1 cm (Table 4) (KRÜGER et al. 1994, LECHNER & BAUMANN 2000, DODEL & HAUPLER 2010, MANSFELD 2010).

Table 4 Characteristics of the SAPOS reference system, modified according to MANSFELD (2010).

Service	*Designation*	*Procedure*	*Transmission*	*Data format*	*Accuracy*
Real-time position service	EPS	RTK	2 m band GSM VHF	RTCM	0,5-3 m
Highly practical. Real-time position service	HEPS	RTK	GSM 2 m band	RTCM	1-2 cm 2-5 cm
Geodat post-proc. service	GPPS	Postprocessing	Email, data carrier	RINEX	1 cm <1 cm

Integrated navigation can be used in the field of *navigation* to increase accuracy and efficiency. This involves the computer-aided processing of data from two or more separately operating navigation systems. The systems record different physical measured values and merge the data, automatically distinguishing between valid and faulty data.

The overall system consists of various sensors, the *reference-dependent* and *reference-providing* systems, the

- Inertial Navigation Systems,
- Map Matching Systems,
- Terrain Contour Matching systems,
- image matching systems (pattern recognition systems), and radio navigation systems, including GNSS (DODEL & HAUPLER 2010).

2.1.3 Geoinformation systems

Geoinformation systems (GIS) are used to process spatial data. The first approaches to modern computer-aided systems were introduced in Canada in the 1960s. Prior to this, geodata had been maintained analogously in register form since the 19th century (BILL 2010).

According to BILL (2010), a GIS is primarily used for computer-aided processing and modeling of spatial problems on an interdisciplinary level. For this purpose, data and information are digitally recorded, modified, managed, analyzed and presented. This can be reduced to a four-component system: Acquisition, Administration, Analysis and Presentation (EVAP). A GIS is therefore to be seen as a superordinate term for a technology or a process, but is also used in everyday language for the software used.

The software products can be divided into seven classes (STEINIGER & WEIBEL 2009, STEINIGER & HUNTER 2013):

- Desktop GIS,
- Spatial database management systems,
- Web Map Server,
- Server GIS,
- Web GIS Clients,
- Mobile GIS,
- Libraries and extensions.

The GIS report published in Germany (HARZER 2014) modifies this classification somewhat; it is divided into the following categories according to functional scope

- GIS (GIS Engine and Client),
- Desktop GIS,
- GIS Viewer

or according to technical characteristics in

- CAD-GIS,
- GIS tray and application,
- Internet GIS,
- GIS enhancement,
- Geodata server,
- Mobile GIS

subdivided. This collection compiles all known manufacturers, providers and products in the field of geoinformation systems on an annual basis. The list currently includes over 1,000 brief descriptions of available individual applications. Fig. 13 shows a selection (BILL 2010, STEINIGER & HUNTER 2013).

Fig. 13 Collection of important GIS applications, modified after BILL (2010), STEINIGER & HUNTER (2013).

Furthermore, GIS applications can be divided into commercial and free products. In the area of free products, there is often talk of open source. However, this is only a partial aspect; open source merely describes the disclosure of the program source code. In connection with free (not necessarily free of charge) availability, these software products are referred to as *FOSS* (Free and Open Source Software) (STEINIGER & HUNTER 2013). This product area in particular has undergone enormous change and continuous further development in recent years, supported by a broad community of supporters. Scientific institutions and faculties in particular are often pursuing open source or free software approaches, as the development possibilities for these applications are almost unlimited. The GNU project defines free software as free if the following four freedoms are met (STALLMAN 2014):

- FreeExecution,
- FreeExamination and adaptation,
- Freeredistribution,
- Freeroom for improvement.

These aspects make it possible for resource-poor areas, municipalities, institutions, corporations and groups in particular to utilize the advantages of computer-aided geoinformation processing (CALDEWEYHER et al. 2006). Increasingly, studies on the use of free and open GIS applications for a wide range of problems are being carried out in all relevant specialist areas (MITASOVA et al. 2005, STEINIGER & HAY 2009, CHEN et al. 2010, NETELER et al. 2012, ZARAGOZÌ et al. 2012, STEINIGER & HUNTER 2012). In the classifications according to STEINIGER & HAY (2009) and HARZER (2014), there are a large number of available applications and collections in the field of FOSS (Fig. 14). In relation to the present work, the desktop GIS applications and internet-based services are of particular importance.

Fig. 14 Free and open GIS applications, as of 2012 (STEINIGER, 2013).

Among desktop applications, the authors currently consider the two systems GRASS (Geographic Resources Analysis Support System) and QGIS (Quantum-GIS) to be at the top in terms of performance, range of functions and distribution (BORK 2008, CHEN et al. 2010). These systems are not limited to the pure, isolated functions as desktop GIS (BILL 2010), but also have interfaces, extensions and import functions, including for the use of network-supported functions such as *web map servers* for the integration of additional layers. Examples of server-supported systems include the geoportals of the federal government, the federal states and other administrative institutions (BKG 2014a). These include, for example, freely accessible high-resolution digital orthophotos with a ground resolution of 40 cm $px^{-1}$and data from the official land register (ALKIS). Comparable applications with different functionalities are WebGIS systems from municipalities and research institutions such as kvwmap (GDI SERVICE 2014) or GeoWiki (FRITZ et al. 2012). A list of other services can be found at the Department of Geoinformation at the University of Rostock (UNIVERSITY OF ROSTOCK 2012) and at the Geoinformatics

Working Group at Osnabrück University of Applied Sciences (OSNABRÜCK UNIVERSITY OF APPLIED SCIENCES 2014). WebGiS applications in the public and popular science sector include applications such as Google Earth, Google Maps, Bing Maps and Open Street Maps (OSM) (CIEPbUCH et al. 2010).

Technically, GiS programs, especially the desktop applications, interpret geoinformation according to the EVAP principle as raster or vector data and associated factual information (attributes) (STEINIGER & WEIBEL 2009, BILL 2010). The raster data model is based on a regular, rectangular or square tessellation of a plane. Each pixel is assigned discrete location information in the form of a two- or three-dimensional coordinate when the geo-information is queried. The homogeneous division is defined by the "mesh width" of the grid, based on the pixel size. Connected attribute information refers to the entire surface of a pixel; in digital elevation models, the average elevation on the pixel surface is usually estimated. In RGB raster maps, red, green and/or blue values are stored as attribute data. Spatially discrete objects are stored in the form of points, lines or polygons as vector geometry with corresponding spatial coordinates. Each object can be assigned one or more pieces of information as attributes. The third layer of the data structure in geoinformation systems is the database layer. This is linked to raster or vector data and stores the attributes in tables. To simplify the database structure, the geoinformation is grouped thematically into layers, which also ensures clarity for the user. For example, individual points in a field survey can be recorded using GNSS systems, processed into complex geometric shapes and superimposed in a layer over a digital orthophoto located in a separate layer. (GOODCHILD 1992, STEINIGER & WEIBEL 2009).

2.2 Precision agriculture

Precise agriculture has been described in detail in countless publications. The following therefore only briefly describes the basics and the development. More in-depth information and studies on user acceptance can be found in MAOHUA (2001), ZHANG et al. (2002), NEMENYI et al. (2003), SEELAN et al. (2003), ADRIAN et al. (2005), NOACK (2006), EHRL & AUERNHAMMER (2007), LAMB et al. (2008), REICHARDT & JÜRGENS (2009), TOZER (2009), NIKKILA et al. (2010), BOGUE (2013), KRUIZE et al. (2013), CUBERO et al. (2014) and MIELE et al. (2014).

The increasing availability of information technology (iT) led to the development of new methods in agriculture in the 1980s and 1990s. Despite all this, the practice at that time had not succeeded in mastering the associated problems such as imbalance between production and consumption in the form of simultaneous overproduction and hunger. AUERNHAMMER (1994) aptly describes the demands on plant production as the basis of human nutrition in his editorial for a special publication on precision agriculture. The greatest challenges lie in the area of information procurement. Agricultural production areas are subject to a high degree of variability and heterogeneity in terms of soil quality, soil fertility and the resulting yield. These factors are regarded as the basis and starting point of *precision farming/agriculture* and date back to antiquity. Small parcelization in the Middle Ages made it possible to individually farm land of varying fertility (STAFFORD 2000). The increasing parcel enlargement in the 20th century increasingly prevented this approach. Therefore, technical methods for *site-specific crop production* were necessary.

COX (2002) divides *precision farming* techniques into a three-stage model:

Data acquisition

- Remote sensing (radar, LIDAR, visible and invisible light, spectroscopy),
- Close-up exploration (photogrammetry, multispectroscopy, NIR, CCD cameras),
- chemical sensors (electronic nose),
- Position data (GNSS).

Data utilization

- Observation and control (machine guidance and control),
- Modeling and data management.

Data exchange

- machine-machine (LBS, ISOBUS, CAN),
- Machine-human (GUI, human-machine interface),
- Human-human (IT, Internet).

The realization of this data stream requires the networking of agricultural implements. One approach is provided by the *agricultural bus system (LBS)* (Fig. 15) and its successor, the *ISOBUS*. These networks are based on the *Controller Area Network (CAN)* and are standardized according to DIN 9684/1 and ISO 11783 (AUERNHAMMER 2000). The ISOBUS is increasingly gaining acceptance as a manufacturer-independent standard for networking and can replace previous proprietary island solutions (BOTTINGER 2010a, ROTHMUND & WODOK 2010).

Initially, precision farming techniques were only used in connection with site-specific management for soil testing, yield mapping, as well as for site-specific fertilization and plant protection measures (AUERNHAMMER et al. 1994, DELCOURT & BAERDEMAEKER 1994, SCHUELLER & WANG 1994, STAFFORD & AMBLER 1994, BASTIAANSSEN et al. 2000).

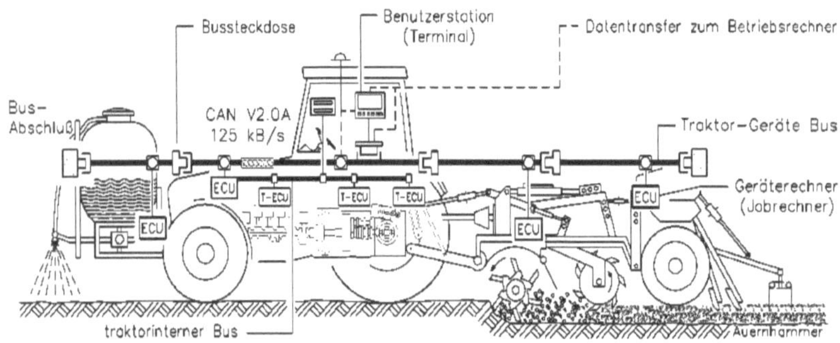

Fig. 15 Schematic structure of the LBS (Vorläufer ISOBUS), modified after AUERNHAMMER (2002).

According to AUERNHAMMER (Fig. 16), precision farming can also provide solutions for *documentation, fleet management* and *field robotics* (AUERNHAMMER 2000, AUERNHAMMER 2001, AUERNHAMMER 2002).

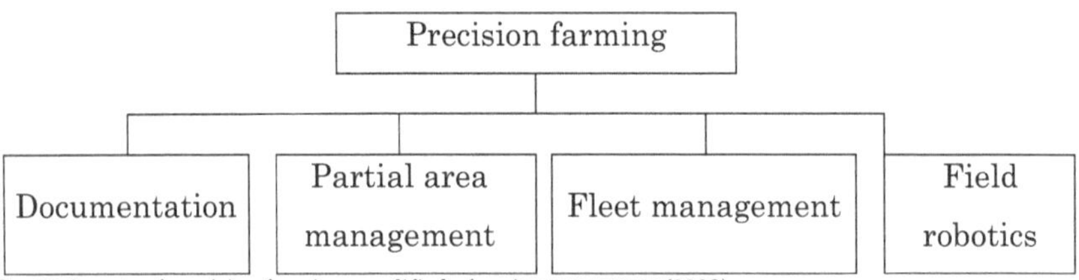

Fig. 16 Aspects of precision farming, modified after AUERNHAMMER (2002).

The individual components of the systems are designed in such a way that the recorded data can be processed, evaluated and prepared directly for documentation (AUERNHAMMER et al. 1999).

One frequently used application of information-based agriculture is *automatic steering systems* or *automatic lane guidance*. Since the mid-90s of the 20th century, more and more satellite-based methods have been developed (LARSEN et al. 1994, STAFFORD & AMBLER 1994). Previously, mainly optical methods, e.g. using lasers, or mechanical methods were known (TILLETT 1991). Automatic track guidance reduces overlapping areas of the working rates and thus increases the flat performance . At the same time, this reduces the environmental impact by saving fuel, reducing CO_2 emissions, reducing the use of pesticides and reducing soil pollution by reducing the number of passes. In the literature, increases in area performance and working speed of up to 15 % are cited (GAN-MOR et al. 2007). The acquisition costs for the necessary technology (GNSS system, automatic steering) can range from €1,300 for simple parallel driving systems to €45,000 for RTK-supported fully automatic steering systems, depending on the degree of automation (BOHRNSEN 2009). However, the use of these is only economically viable in inter-company use (KÜBLER 2010). To assess the purely monetary advantages of automatic lane guidance, a differentiation must be made between work processes. In pure driving operation, for example in soil cultivation, the expected cost savings due to reduced overlapping are only in the range of €0.09-0.25 ha^{-1}, depending on the working width (BATTE & EHSANI 2006, BOHRNSEN 2009). As soon as other inputs such as pesticides or fertilizers are applied, the cost reduction due to savings in funds is significantly higher. KÜBLER (2010) reports a cost advantage of €20.12 ha^{-1}through the use of automatic steering systems based on a farm size of 1,600 ha. This saves a total of €25,184 a^{-1}. Studies in vegetable cultivation have shown potential savings of up to 60% in process costs through the use of satellite-supported hoeing technology for weed control compared to conventional methods. At the same time, fuel and other operating resources are saved, which contributes to improving the CO_2 balance. The working time requirement was reduced by up to 70 % due to larger working widths (SCHWARZ & HEGE 2014).

In order to be able to characterize the positional accuracy when using automatic steering systems, the errors that occur are broken down in relation to the direction of travel. On the one hand, deviations transverse to the direction of travel (transverse) and on the other hand along the path (longitudinal) are assessed. When looking at the resulting position errors, PEREZ-RUIZ et al. (2012) found RMs errors (Root Mean square) of 23.9 mm (transverse) and 26.7 mm (longitudinal) when planting tomato seedlings at a driving speed of 1.6 km h^{-1}. The total error amounted to 55.8 mm on average. GAN-MOR et al. (2007) investigated the deviations of implements depending on the distance to the rear axle in agricultural tractors. The position error of the tractor rear axle when driving on uneven terrain was 36.4 mm on average at a speed of 4 km h^{-1}. NORREMARK et al. (2007) found maximum deviations of 37.3 mm when placing sugar beet seed. Iongitudinal errors tended to be larger in these cases. All authors used guidance systems with vehicle tilt compensation to correct the track. NORREMARK et al. (2007) investigated their influence and found significantly larger position errors when no inclination compensation was used. Depending on the direction of travel, 1.6-59 times greater deviations were recorded without tilt compensation.

In addition to purely satellite-based navigation systems, the use of other sensor technology is also possible. Systems with mechanical scanning of an inventory are available here. Increasingly, systems with contactless, optical or acoustic tracking have been developed and offered over the last two decades. For example, imaging methods with CCD cameras (Charged Coupled Device), 3D image processing, laser scanners or ultrasonic sensors are used (KEICHER & SEUFERT 2000, KEICHER 2002, BOTTINGER 2010b).

2.3 Robotics

In the middle of the 21st century, the world population is expected to reach around nine billion people. To adequately feed this number, global food production would have to be increased by 70-100%, which will lead to competition for land, water and energy. One direct competitor to food production is, for example, the production of biofuels (GODFRAY et al. 2010). The further automation of agriculture can, among other things, provide solutions (WUNDER et al. 2012).

Like AUERNHAMMER (2001), RUCKELSHAUSEN (2010) also sees field robotics as the next stage of automation in agriculture. He presents several options for fleet management in field cultivation:

- Combination of manned and unmanned vehicles of existing concepts (semi-autonomous fleets).
- Purely unmanned vehicles of existing concepts (autonomous fleets).

- Combination of manned vehicles of existing concepts and unmanned vehicles of new concepts (semi-autonomous fleets with field robots).
- Unmanned vehicles of new concepts (autonomous field robots - schwărme).

For example, standard tractors with automatic steering systems are assumed to be existing concepts. The authors see the transition to fully autonomous systems as a gradual process.

One example of this is the *electronic drawbar* concept. This involves the virtual coupling of standard tractors equipped with automatic steering. This link is realized via a radio link and is equipped with various hardware and software modules. It works according to the *master-slave principle*, whereby the master tractor is guided by a driver, but can also be steered by means of automatic track guidance (NOACK et al. 2010).

The problem of large machines and the resulting soil compaction is often pointed out (AUERNHAMMER 2001, BLACKMORE et al. 2009). For this and other reasons, smaller, lighter and more flexible machines are used in field robotics. This increases maneuverability and facilitates the processing of awkwardly cut fields. The scaling effects of using large working devices are compensated for with smaller autonomous machines by longer operating times (24 hours), the use of many smaller machines and more efficient use (e.g. stock detection) (BLACKMORE et al. 2009, GRIEPENTROG 2010).

VOGELE (2010) summarizes the requirements for a sophisticated mobile agricultural robot as follows:

- Off-road capability and robustness,
- Energy autonomy,
- Orientation ability in the terrain,
- Obstacle detection,
- Changing routes and tasks when obstacles occur,
- Efficient cooperation with humans and other robots,
- Cognitive and manipulative skills for autonomously solving complex tasks.

The Field Robot Event has been held every year since 2003. Here, student teams from various European faculties compete against each other in several disciplines and demonstrate the current state of research in the field of field robotics (MARTINET & BLAAUW 2014).

In addition, various prototypes for use in arable farming and special crops have been developed in a number of research projects over the past few years. The area of robotics in dairy farming in the form of milking robots is deliberately excluded here. Some field robotics projects are mentioned as examples.

As early as the end of the 1980s, studies were carried out in Italy on the full mechanization of pruning using a tractor-assisted pruning robot (VANNUCCI et al. 1990). Since 2008, the US company Vision Robotics has been conducting trials to automate pruning using 3D image recognition (BLOSS 2014, VISION ROBOTICS 2015). Since 2010, a working group in New Zealand has been trying to transfer human cognitive processes in pruning into rules for artificial intelligence for use in a robot (SAXTON et al. 2014).

In 1996, a model for harvesting fruit and vegetable crops was described in Japan (KONDO et al. 1996). Also in Japan, a prototype of a harvesting robot for cherries with 3D image recognition and a harvesting organ with four degrees of freedom was produced and tested (TANIGAKI et al. 2008). Another study dealt with the use of a harvesting system for strawberries in greenhouses with subsequent placement in trays. The system required around three times the time compared to humans (HAYASHI et al. 2010).

The U-Go robot is an attempt to establish an electrically powered multi-functional platform for viticulture. The vehicle can either be remote-controlled with a joystick, guided semi-autonomously along rows of vines using a laser scanner or navigated fully autonomously with satellite support. A sprayer for plant protection application is currently available (LONGO et al. 2010, LONGO et al. 2012).

The VineRobot research project is currently investigating the automation of non-invasive scoring techniques in viticulture. The aim is to use a ground-guided vehicle to record the parameters of yield, vegetative growth, water status and berry constituents (VINE ROBOT 2015).

A Chinese working group has developed and tested a robot for apple harvesting. The tracked platform is navigated via GNSS. Imaging methods are used to detect apples. On average, 15.4 s were required per fruit with a success rate of 77 % (DE-AN et al. 2011, JI et al. 2012).

Initial simulations and modeling were carried out in Italy to automate the harvesting of artichokes. The authors have investigated possibilities for mechanical harvesting, object recognition and impact force (LONGO & MUSCATO 2013).

The *BoniRob* (Fig. 17), a cooperation project between scientific and industrial project partners, is particularly noteworthy in the field of automated phanotyping, for example the assessment of plant populations. A platform was developed to accommodate various *applications* for scoring, soil analysis or weed control (BANGERT et al. 2013). It is driven by wheel hub motors with a variable steering concept (RUCKELSHAUSEN et al. 2009). The concept of *sensor fusion* is used to navigate the vehicle, i.e. the data from several sensors (LIDAR, inertial, RTK-GPS) are combined in the multi-layer navigation system (BIBER et al. 2012).

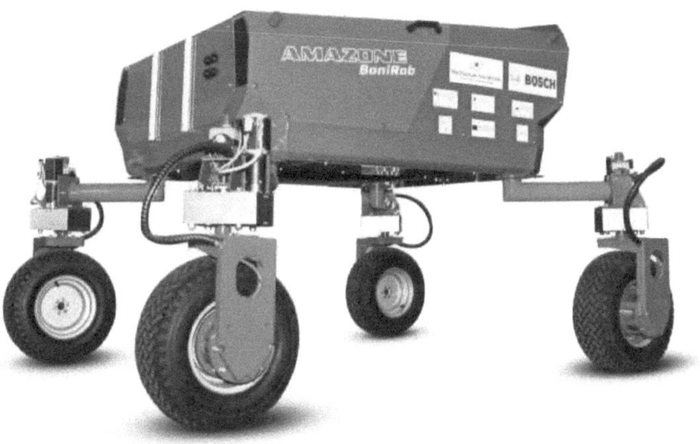

Fig. 17 Phanotyping platform BoniRob V2 (AMAZONEN-WERKE 2011).

In the second generation, the system should be usable in the same way as tractor-mounted răt combinations. The modular application mounting enables multifunctional use. Fig. 18 shows the process of weed control as a cooperative process involving the BoniRob V2 and a human operator for the semi-automatic detection of weeds (RemoteFarming.1a). In the next stage of development, this process is to be further automated using adaptive algorithms (RemoteFarming.1b) in order to relieve the operator. Complete automation is not currently being pursued (BANGERT et al. 2013).

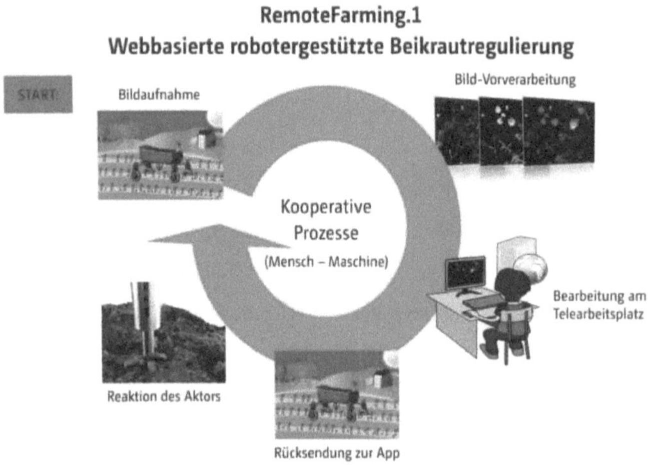

Fig. 18 Process sequence RemoteFarming.1a using the BoniRob V2 platform and a teleworking station for semi-automatic weed control (AMAZONEN-WERKE 2011, BANGERT et al. 2013).

According to the authors, the *BoniRob* concept is one of the systems closest to commercial use (BANGERT et al. 2013).

An interdisciplinary research project currently under development can draw on the experience gained from the *BoniRob* project (LINZ et al. 2014b). *The elWObot* platform for the application of plant protection agents in viticulture and fruit growing uses approaches from sensor fusion, navigation concepts and control by means of a *Robot Operating System (ROS)* (LINZ et al. 2014a). The *GAZEBO* simulation environment was used to develop the navigation algorithms in this and some of the above-mentioned projects (RUCKELSHAUSEN et al. 2009, EMMI et al. 2013, LINZ et al. 2014a, LINZ et al. 2014b).

2.4 Phanotyping and remote sensing

Developments in the field of electronically supported precision agriculture are constantly revealing new applications for these technologies. In general, development and application advances can always be seen in the field of flatland agriculture; new methods usually spread first in this field. Techniques from the chapters 2.2 and 2.3 are used to assess crop parameters such as soil quality and condition or plant characteristics. Particularly in the observation of plants and the recording of phenotypic characteristics, there is a desire to be able to carry out the assessment non-invasively, i.e. without injury or damage. Non-contact methods have been used here for several years. These can be classified as near or remote sensing. A variety of systems and methods are now available for a number of agricultural crops. There are several comprehensive overviews of available applications in the literature (BASTIAANSSEN et al. 2000, NICOLAÏ et al. 2007, LEE et al. 2010, RUIZ-ALTISENT et al. 2010, SANKARAN et al. 2010, AQEEL-UR-REHMAN et al. 2014, DEERY et al. 2014).

In scientific usage, the term *remote sensing* is often used as an umbrella term, which is the direct translation of *remote sensing*. The Springer-Gabler Wirtschaftslexikon provides a definition (NEUMAIR & HAAS):

"Method for observing the earth's surface, the sea surface and the atmosphere from airplanes or spacecraft (satellites), which use the electromagnetic radiation emitted by the objects to obtain information (remote sensing)."

In addition, remote sensing sensors can also be guided over the earth's surface using drones or balloons (ALBERTZ 2009). The use of drones in particular (Fig. 19) has increased significantly in recent years and can be implemented cost-effectively compared to conventional technologies. However, the legal requirements for operating these unmanned aerial vehicles (UAVs) are not insignificant (GEO-KONZEPT 2014).

Fig. 19 Drone (Octocopter) from the company geo-konzept during a demonstration at the Institute of Viticulture at Geisenheim University.

The Octocopter from geo-konzept can be loaded with a payload of approx. 1.5 kg. On the one hand, this can be a commercially available system camera, from whose images 3D terrain images with a ground resolution of 10 mm px^{-1}can be created (Fig. 20). On the other hand, the system can be equipped with a multispectral camera (GEO-KONZEPT 2014).

Fig. 20 Drone (Octocopter) from the company geo-konzept during a demonstration equipped with a standard Sony system camera for generating 3D gel images.

The above-mentioned image types (RGB, multispectral) are examples of passive remote sensing systems in addition to thermal images. Here, the reflection of solar radiation or radiation emitted from the surface is used. Active systems work by emitting microwaves (radar) or other types of waves (laser) (VOSS 2012). Remote sensing can also be divided into the fields of *photogrammetry* and *satellite geodesy* (ALBERTZ 2009).

Remote sensing technologies have also been developed, adapted and used for viticulture for several years. At the beginning of the 2000s, a number of studies investigated the potential uses of available satellite technologies (Landsat, SPOT, IKONOS) and aerial photography techniques as well as other non-contact measurement technologies (HALL et al. 2002). On the one hand, soil investigation methods with electromagnetic sensors can be used to assess the water supply via the electrical conductivity of the soil (LAMB et al. 2002). These investigations were also compared with imaging methods in order to verify their informative value (HALL et al. 2002). On the other hand, in addition to the good results for characterizing the soil and its zonation, the image-based methods can also be used to assess and describe the covering vegetation (LAMB et al. 2002). Images taken with multispectral cameras can be used to calculate various indices, such as the

Normalized Difference Vegetation Index (NDVI), which can be used as an indicator of plant growth and biomass formation (JOHNSON et al. 2001, HALL et al. 2002, HALL et al. 2003, ARNÓ et al. 2009). The photosynthetically active biomass can be used to make statements about secondary fungal constituents such as phenols. This makes it possible to investigate the potential phenol content at maturity during the vegetation phase (LAMB et al. 2004).

In addition to airborne remote sensing methods, these techniques can also be used on the ground. In the case of viticulture, the sensors are guided along the foliage wall. RGB and multispectral images can also be used here (FERNANDEZ et al. 2013, KICHERER et al. 2013).

3 Material and methods

This paper focuses primarily on the development and construction of the automated PHENObot scoring platform as part of the PHENOvines research project. The scoring methods and results obtained with it are part of the publications of colleagues at the JKI, Geilweilerhof site, and are only dealt with in passing.

3.1 Vehicle development

In the cooperation agreement based on the project application, it was agreed that the development and production of the carrier vehicle should be as cost-efficient and resource-saving as possible. First, a requirements profile was drawn up for the platform:

- Compact size and low center of gravity,
- Payload at least 250 kg,
- Low-vibration drive,
- Travel speed 2-4 km h^{-1}.

In order to minimize development costs, it made sense to use as many series components as possible. Taking the requirements profile into account, a survey of the vehicle concepts and vehicle parts available on the market was carried out. The results of this evaluation and the course of further development work are described in Chapter 4.

3.2 Constructive work

The design and execution were carried out at the Institute of Technology at Geisenheim University. The CAD software SolidWorks 2011 SP 5.0 was used for planning and design. The vehicle was created with all components as a 3D model. The 2D production drawings created from this could be used directly for production. Some workpieces had to be made from high-strength aluminum; an external metalworking company was commissioned to produce them.

3.3 Navigation

Due to the requirements for the scoring platform, the robot was not realized as a fully autonomous system. In principle, the system should automate the previously manual scoring of the vines as far as possible. The individual breeding lines or clones are arranged as individual vines or in small groups in rows of vines. There is a coding system that assigns a unique identification number to each vine. All vines are georeferenced and positioned with unique coordinates so that they can be found again. In order to keep the overall system lean and simple, an autonomous vine recognition mode was therefore dispensed with. The assignment and control is carried out purely on the basis of the pole coordinates by planning travel paths.

GIS programs were used for path planning when navigating the robot and evaluating position-related data (see section 2.1.3).

The position data of the vines in JKI facilities were available as Cartesian coordinates (3D) in the ETRS89 coordinate reference system in the UTM projection. The *high* component was not taken into account during positioning, as the vehicle's control application only worked with two-dimensional coordinates. This position data was supplemented with additional information during development and transferred as a database.

In addition to the unique coordinates `(easting, northing)`, this database contains an identification number `(stockID)` assigned to the stock in the format

`Parcel number-Vine number-Stock number`

(e.g. `23-001-001`), the bending direction of the bending beam and the identification of the start and end points of the line. The `stockID` is used for the

Image acquisition is assigned to each image and the transferred coordinates, making it easier to assign the images. More details on automated job output in chapter 4.6

3.4 Software

The entire control application and the GIS programs were executed under the Linux-based operating system Kubuntu 12.04.4 LTS. All libraries of the vehicle control system were available for this system, and the other applications were also adapted to it for standardization.

The freely available open source software GRASS GIS version 6.4.4 was used to create the paths. This is very well suited for command line-oriented work and has an enormous range of functions for processing and transforming geodata. The available stock coordinates were processed and imported into GRASS GIS in an ASCII text file. The import was automated using a shell script (bash), whereby some vector shifts and calculations were carried out. In the final step of the script, the data was exported to another ASCII text file, which was transferred to the navigation computer of the PHE NObot as a so-called *job list.*

During the processing of this *job list*, the control application creates several log files that can be used for evaluation and function checks. For this purpose, Quantum GIS (QGIS) version 2.6 was used as a further GIS program due to its good graphical display options. QGIS offers a large number of additional plug-ins for in-depth analysis of geodata; the additional modules for creating so-called heat maps and the output of azimuthal polylines were used here. Furthermore, the available digital orthophotos and the official real estate maps of the state-specific geoportals (Geoportal Rheinland-Pfalz and Geoportal Hessen) can be integrated as Web Map Service (WMS) layers (Fig. 21).

Fig. 21 Screenshot of trial plot P23 at JKI Geilweilerhof, exported from QGIS. The colored dots mark the stopping points for image acquisition. (digital orthophoto from Geoportal RLP 2014).

The manufacturer of the drive components of the carrier vehicle supplied libraries in the object-oriented programming language C++ for control purposes. The control application was written in the integrated development environment (IDE) Qt

Creator 4 (Fig. 22). The C++ compiler g++ from the GNU Compiler Collection (GCC) was used as the compiler. The entire program development again took place under Kubuntu 12.04.4 LTS and Debian 7 (Wheezy).

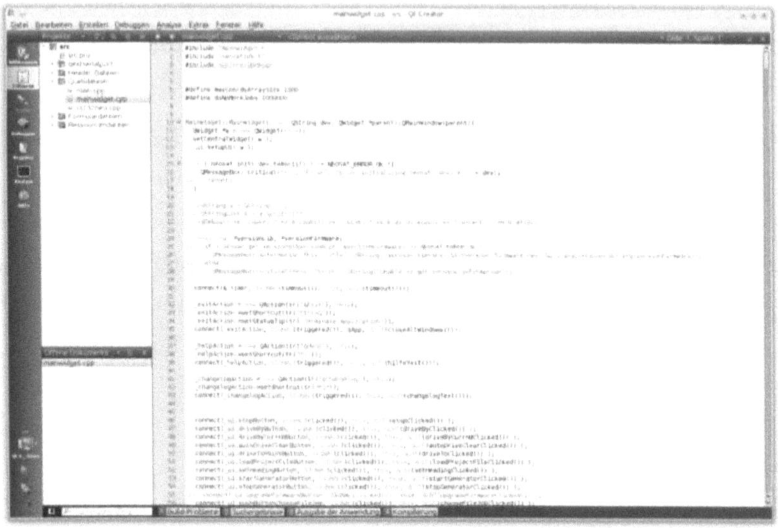

Fig. 22 Screenshot of Qt Creator 4 with source code of the control application.

3.5 Test areas

The test runs in the field took place both on test plots at Geisenheim University (Fig. 23), in the federal state of Hesse, and at the JKI, Geilweilerhof site (Fig. 24), in the federal state of Rhineland-Palatinate. After completion of the project, the PHENObot will remain at the JKI in Siebeldingen and be used in the plants there. These are designed as classic trellis-trained vineyards with an aisle width of 1.90-2.00 m and a cane spacing of 1.00-1.20 m. The PHENObot is to be adapted to these conditions. The dimensions of the PHENObot are to be adapted to these conditions. During the development phase, tests were carried out in Geisenheim on areas belonging to the university. The tests took place primarily in three areas. An area of around 1,100 $m^2$was available as a test area in the courtyard of the Institute of Technology. The construction and software tests were carried out here. The area of around 400 m2 at the institute's greenhouse was mainly used for testing and presentation purposes in the second year of the trial. This was a woody plant trial of the Landesbetrieb Landwirtschaft Hessen (LLH), which is laid out in rows equivalent to vine alleys. In the final project phase, a plot of the Institute of Vine Breeding with eight rows of vines was measured using the RTK-GPS system (Trimble) of the Institute of Technology. Here the coordinates of the individual vines, the end anchoring and the access route were recorded according to the defined surveying style (Section 3.6.1). The structure, alignment and condition of the site were comparable to the test sites in Siebeldingen, which ensured the comparability of the results obtained in each case.

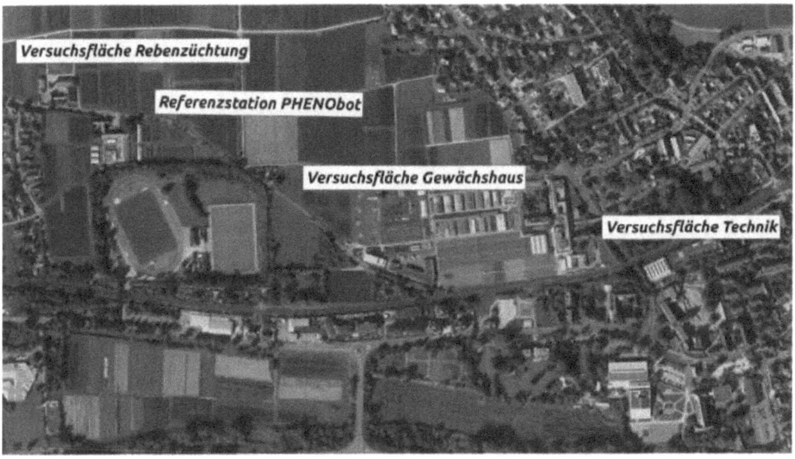

Fig. 23 Experimental fields (red areas) in Geisenheim (digital orthophoto Geoportal RLP 2014).

The test vineyard was planted in 2011 with a row spacing of 1.80 m and a cane spacing of 1.30 m. It is a mixture of varieties, including St. Laurent, Traminer, Pinot Noir and Pinot Gris. Borner was used as the rootstock (KONRAD 2015). The average row length of the eight rows is 65 meters. Each row was measured in the expected direction of travel, i.e. the survey began on the south side with the eastern row of vines and was carried out in a northerly direction. At the end of the row, the survey was continued with the western row in a southerly direction. The coordinates of the measuring points were given a code:

`RZR[`Line number`]S[Stock number].`

It began with the end anchorage (floor number 0). On the eastern side, a farm road directly adjoined the plot, the edge line was numbered 34. This resulted in a code range of `RZR34S0-RZR34S47` for this row. This coding was passed on when the *job lists* were created, so that a coordinate could be clearly assigned to the respective stick or measurement position (see Chapter 3.8) during the evaluation.

Fig. 24 Experimental site JKI Geilweilerhof (red area) (digital orthophoto Geoportal RLP 2014).

Trials were carried out or prepared in four individual plots at the JKI. In the first trial year 2013, test runs were carried out in plots 19 and 51. In addition, trials were carried out in plot 23 during the vegetation phase in 2014. In this plot 23, the final application test on the operational and performance capability of the entire image acquisition system took place in September 2014. The gas sensor width of the system is 2.00 m. With a cane spacing of 1.10 m, this results in a standing space per cane of 2.10 m^2Plot 53 was used to test the automated *job list creation.*

3.6 Measurement technology

The measurement of the pole coordinates and the determination of the relative and absolute accuracy (see section 3.8) was carried out using GNSS surveying systems. Construction industry systems from Trimble, which consist of three main components, were used at both test sites:

- RTK reference station (base unit),
- Rover (mobile unit),
- Control unit.

A UHF radio link in the 433 MHz frequency range is used to connect the reference station and the rover. The configuration was carried out with the aid of the manufacturer's technical documentation (TRIMBLE NAVIGATION LIMITED 1998, TRIMBLE NAVIGATION LIMITED 2001).

3.6.1 Trimble Geisenheim surveying system

The Trimble 4400 GPS receiver was used as the base station. This receives the signals from the GPS satellites on the L1 and L2 bands. The Trimble 4700 GPS receiver acted as a rover (also L1/L2). Identical GPS antennas of the Trimble Micro-Centered L1/L2 model were connected to both receivers. The Trimble 4700 rover had an internal modem, while the Trimble 4400 base station used the external Trimble Trimtalk 450S radio modem. The radio connection was established at 433.4 MHz in the CMR plus transmission format. Both modems were operated with tuned radio antennas of the type Trimble 24253-44.

The Trimble Survey Controller 1 (TSC1) was used for configuration and operation. Individual projects (named after the respective test variants) of the same configuration were created for the measurements. The coordinate system used was UTM in zone 32 North with the datum format WGS84. The large half-axis was 6,378,137,000 m, the offset 298,257223563 (dimensionless). These were automatic specifications from the TSC1 library. A comprehensive *PHENObot* measurement style was created during project creation. This defines all parameters and settings for the operation and execution of a survey, including settings for the rover, the base, the radio protocol, the units used, the number of individual measurements and coordinate formats. The datum format ETRS89 with the reference ellipsoid GRS80, which is now used in official surveying, was not included in the configuration of the TSC1. A calibration to the ETRS89 spatial reference system was carried out with the help of fixed points from the Hessian State Office for Soil Management and Geoinformation. For this purpose, the Trimble reference station was positioned on the fixed point TP 601380/04 (Tab. 5; GPS-compatible point on Geisenheim Cathedral Square) (Fig. 25 and Fig. 26) and the official coordinates were entered in the configuration according to the ETRS89/UTM coordinate reference system.

Fig. 25 Setting up the reference station on the position fixed point TP 601380/04 Geisenheimer Dom- platz.

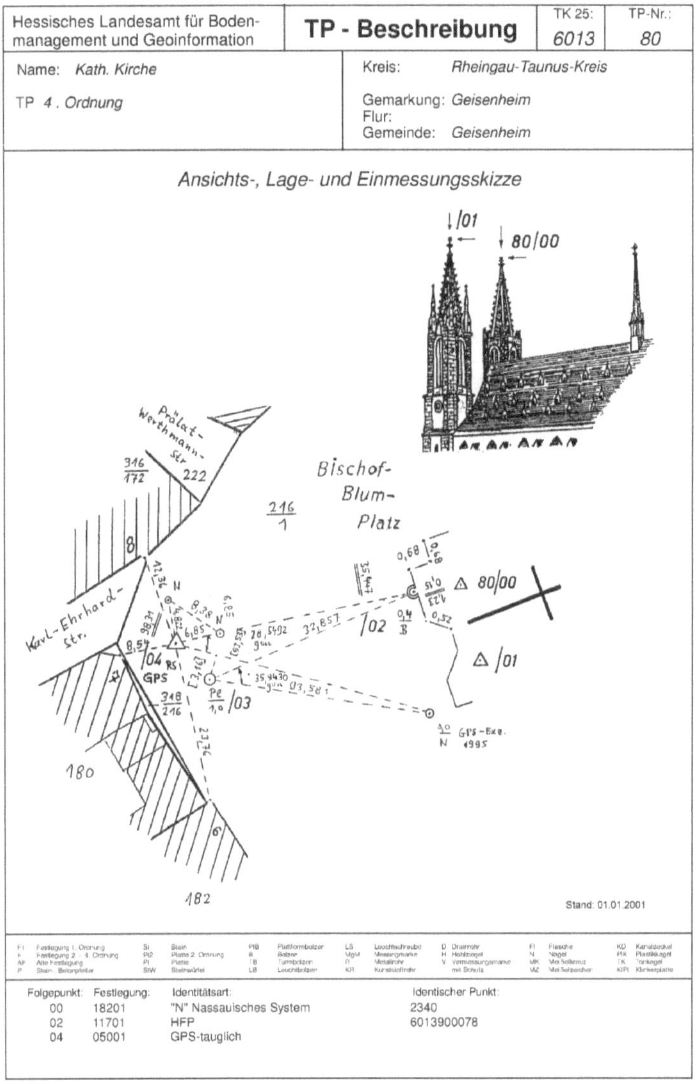

Hessisches Landesamt für Bodenmanagement und Geoinformation	TP - Beschreibung	TK 25: 6013	TP-Nr.: 80

Name: *Kath. Kirche*

TP *4*. *Ordnung*

Kreis: *Rheingau-Taunus-Kreis*

Gemarkung: *Geisenheim*
Flur:
Gemeinde: *Geisenheim*

Ansichts-, Lage- und Einmessungsskizze

Folgepunkt:	Festlegung:	Identitätsart:	Identischer Punkt:
00	18201	"N" Nassauisches System	2340
02	11701	HFP	6013900078
04	05001	GPS-tauglich	

Fig. 26 Location fixed point description TP 601380/04 Geisenheimer Domplatz, HESSISCHES LANDESAMT FÜR BODENMANAGEMENT UND GEOINFORMATION *(2014).*

This calibration was used to measure the intended position of the reference station on the roof of the Institute of Technology. For this purpose, the GPS antenna was mounted on a telescopic mount (Fig. 27) and connected to the Trimble 4700 rover.

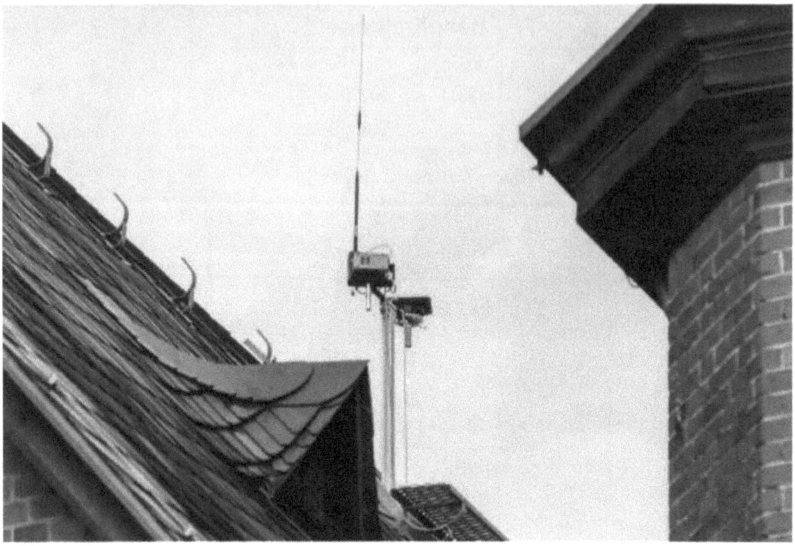

Fig. 27 GPS antenna (right) and radio modem (left) of the reference station (Trimble) on the roof of the Institute of Technology, Hochschule Geisenheim University.

In addition, the radio antenna, a DC voltage of 12 volts and the TSC1 were connected to the rover (Fig. 28). A survey was started in the TSC1 menu to record point coordinates. The preconfigured *PHENObot* surveying style was stored. The connection between the rover and the base station was established. According to the manufacturer's specifications, at least five available satellites are required to initialize the measurement. This results in the so-called RTK fix, which is the most accurate measurement method with this system. The *fixed point* measurement method was selected. This involved taking 180 individual measurements at a frequency of 1 Hz over a measurement period of three minutes, from which the TSC1 calculated a three-dimensional position in the UTM grid. Subsequently, the Trimble 4400 base receiver was connected to the antennas of the reference station and the newly measured coordinates were communicated to it via the TSC1. The base station was thus virtually at a position (Table 5) in the ETRS89/UTM coordinate reference system and sent the connected rover a correction signal corrected for the offset between WGS84 and ETRS89.

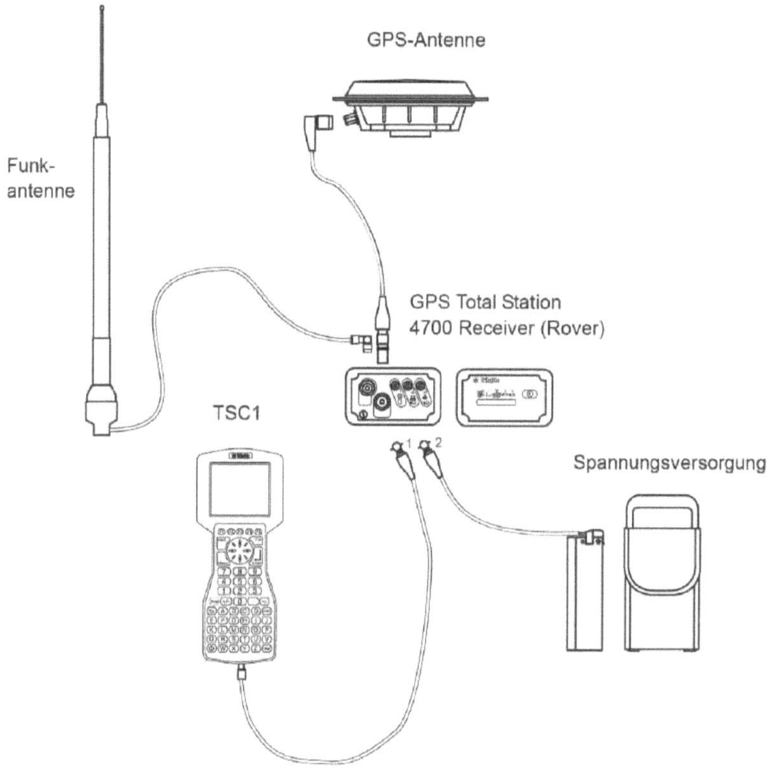

Fig. 28 Schematic diagram of real-time kinematic measurement Rover, modified after TRIMBLE *(2001).*

Tab. 5 Three-dimensional coordinates calibration Trimble reference station Geisenheim; Coordina- tion reference system: ETRS89/UTM; High WGS84 = High above sea level + undulation (48.34 m).

Point name	***Right value [m]***	***High value [m]***	***High WGS84 [m]***
TP 601380/4	425907,829	5537209,922	136,301
Reference station Institute of Technology	425644,575	5537320,944	163,884

The Trimble surveying system was used for the RTK measurements during all tests for the relative and absolute accuracies, as well as the measurement of the floor coordinates, according to the scheme in Fig. 28 in the *PHENObot* surveying style. The points were recorded using the *topographic point* measurement method with three individual measurements at a frequency of 1 Hz with a minimum occupancy time of three seconds and a horizontal accuracy threshold of 0.015 m. If this threshold was exceeded, the measurement paused until it fell below it. When measuring points with the surveying rod, care had to be taken to ensure that the GPS antenna was always vertically aligned above the measuring point (Fig. 29).

There were two circular spirit levels on the rod for this purpose.

In addition, so-called *continuous measurements* were carried out with the GPS system during some test drives. On the one hand, points were recorded continuously at a frequency of 1 Hz, and on the other hand at a resulting position change of $\Delta\chi = 0.1$ m.

The TSC1 operating unit worked with a memory card in *PC card* format, size 128 MB. All projects were created on this card and saved the recorded points on it in a proprietary file format. The point information was exported to ASCII text files (csv file) using the export function of the TSC1 and read from the memory card for further evaluation:

- Right value [m],
- High value [m],
- High (corrected) [m],
- Point name,
- Name code.

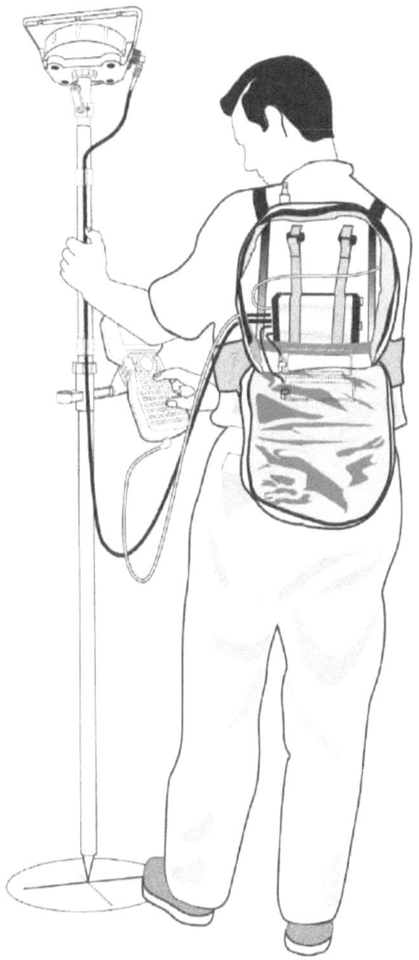

Fig. 29 Perpendicular alignment of the GNSS antenna above the measuring point during a real-time kinematic measurement, modified according to TRIMBLE (1998).

3.6.2 Trimble Siebeldingen surveying system

Schematically, the JKI's GPS system was comparable, but of more recent construction. Two identical Trimble SPS852 GNSS receivers, each with a built-in radio modem, were used for the rover and the reference station. Two Trimble Zephyr Model 2 antennas were used as GNSS antennas. The radio link was established using the CMR+ protocol with a frequency of 433.0 MHz. This system was mainly used to identify the recorded image data with geoinformation. In addition, the vehicle's reference station was calibrated once using this system and the position stored in the firmware. The geoinformation was also evaluated in order to check the relative and absolute accuracy when driving in Siebeldingen. This was done by exporting the data to ASCII text files.

3.7 Documentation

This point is divided into two sections: The creation of protocol files during the use of the test carrier and the preparation of the operating manual.

3.7.1 Application protocols

Log files were created in the control application to evaluate the test drives and check the function of all hardware and software components. When the log function is activated, certain parameters are output in an ASCII text file. Four different outputs or functions were provided:

- `logAll`,
- `logJob`,
- `logPos`,
- Command line protocol.

Various vehicle and navigation parameters and coordinates were output for the `logAll`, `logJob` and `logPos` functions:

- `x` (relative horizontal distance to the RTK base in meters),
- `y` (relative vertical distance to the RTK base in meters),
- `easting` (easting as UTM coordinate),
- `northing` (north value as UTM coordinate),
- `heading` (orientation vehicle right-pointing north in decimal degreesß),
- `GPS1stat` (GPS antenna 1 status message),
- `GPS2stat` (GPS antenna 2 status message),
- `GPS1numSat` (number of GPS1 satellites received),
- `GPS2numSat` (number of GPS2 satellites received),
- `autodrive` (status message driving mode),

- `time` (time specification CET or CEST in hh:mm:ss),
- `StockID` (unique character string for a point or vine).

The parameters were separated by a concatenation character (vertical bar |), the ASCII text file had the extension `*.txt`.

logAll

This function made it possible to log the parameters `x, y, easting, northing, heading, GPS1stat, GPS2stat, GPS1numSat, GPS2numSat, autodrive` and `time` at 10 Hz independently of the execution of a work order.

logJob

This function made it possible to log the parameters `x, y, easting, northing, heading, GPS1stat, GPS2stat, GPS1numSat, GPS2numSat, autodrive` and `time` at 10 Hz during the execution of a work order.

logPos

This function made it possible to log the parameters `easting, northing, heading` and `StockID` for insertion into a *job* list as a separate work order. A single `logPos file` was created for each *job list* started; the parameters were written to a new line of this file each time the function was executed.

Command line protocol

The control application was designed as a command line program with a graphical user interface (GUI). When the application was executed, a continuous text output of the executed functions followed at the command line level, which were sequentially saved in an ASCII text file (file extension `*.log`).

3.7.2 Manual

As a machine for use in research, the PHENObot is excluded from these requirements in accordance with the new version of the Machinery Directive 2006/42/EC §1 sentence 1h. Nevertheless, an attempt was made to adhere to the principles of this directive when designing and creating the operating instructions (LENZ & OTTO 2013). The word processing program Microsoft Office Word 2010 was used to write the manual. The manual is included in the electronic appendix.

3.8 Test drives

In principle, a distinction is made between test and measurement runs. The former were used to test design developments, changes and adaptations, including in the software and navigation sector. The measurement runs included the quantification of some process and accuracy parameters.

3.8.1 Test drives Geisenheim

These were mainly carried out close to the Institute of Technology (see section 3.5: Farm Technology, Flat Greenhouse, Vine Breeding). No test protocol was defined here; an individual procedure was used depending on the question. In all tests with GPS-supported operation, the RTK base station was generally commissioned first to provide the correction signal.

3.8.2 Measuring trips Geisenheim

The measurements to determine the relative and absolute accuracy of the navigation solution were carried out on the "vine cultivation" area.

The log functions `logJob` and `logPos` were used to determine the position of the vehicle. The coordinates of the position-determining GPS antenna-1 were recorded via the control application. A virtual point located at the vehicle's pivot point was used as the navigation point. This could be reconstructed tri- gonometrically from the raw data of the logs. The detailed structure of the vehicle configuration can be found in Chapter 4.1; the correction calculation is listed below. It should be noted in advance that the position of the GPS antenna 1 deviates by 350 mm from the pivot point in the x-direction and 800 mm in the y-direction. The following equations only use the variables x and y. The resulting navigation point position in the Cartesian coordinate system depends on the vehicle orientation. A two-stage angle calculation must be carried out to determine the displacement factors Δx and Δy.

$$\propto_d = tan^{-1}\left[\frac{x}{y}\right] \qquad \text{Gl. 3-1}$$

Eq. 3-1 is used to calculate the angle κ of the triangle resulting from x and y as the opposite and anchor sides.

$$\propto = 180° - heading + \propto_d \quad \text{Gl. 3-2}$$

Eq. 3-2 gives the angle for the orientation-dependent position shift of x and y. The value `heading` reflects the vehicle orientation κ.

$$\Delta x = cos \propto \cdot \sqrt{x^2 + y^2} \quad \text{Gl. 3-3}$$

$$\Delta y = sin \propto \cdot \sqrt{x^2 + y^2} \quad \text{Gl. 3-4}$$

Eq. 3-3 and Eq. 3-4 calculate the displacement of x and y. The virtual navigation point *P0* is added to the initial coordinates:

$$P_0 = [x + \Delta x \mid y + \Delta y] \quad \text{Gl. 3-5}$$

To determine the absolute accuracy, this point *P0* was first compared with the floor coordinate *S0* to be approached by calculating the difference between the individual vectors.

The deviation was differentiated in the transverse and longitudinal direction as root mean square (RMS).

From this, the average deviation of the actual and target position could be calculated using only the vehicle's own data. In order to validate this data, a secondary measurement was carried out using the GPS system from (Trimble) described in Chapter 3.6.1. The GPS antenna was mounted at a defined point in the center of the vehicle and connected to the rover. The measurement was repeated three times with the identical configuration used to measure the floor coordinates (project and measurement style *PHENObot*).

The individual values of the three repetitions were used to determine the relative accuracy by calculating the difference between the vectors. The transverse offset referred to the perpendicular deviation of the position to the lane axis. The deviation in the longitudinal direction was calculated as the difference along the direction of travel. The calculation was based on the differences of the right and up values from the log files together with the recorded vehicle orientation x:

$$\Delta transversal = \frac{\Delta x}{\cos \alpha} \qquad \text{Gl. 3-6}$$

$$\Delta longitudinal = \frac{\Delta y}{\cos(180-\alpha)} \qquad \text{Gl. 3-7}$$

This was done using the vehicle's own data and the secondary data.

The fully developed foliage in the rows of vines was expected to have an impact on the reception quality of the GPS antennas. One measure of the quality of reception can be the number of available satellites. Using the log function of the control application, the number of available satellites for both existing GPS antennas of the vehicle with fixed position assignment could be recorded with /=10 Hz. The arrangement of these antennas was made variable via a shifting mechanism in order to be able to realize positioning at different heights. Measurements of availability were taken in the lowest setting (`ant_down`) (height above ground: 1,280 mm, approximately in the middle of the foliage wall) and the highest setting (`ant_up`) (height above ground: 2,000 mm, upper end of the foliage wall) in triplicate. Using the GIS software QGIS, the recorded values could be visualized on a choropleth map or a so-called heat map. For this purpose, the feature "number of satellites" in the attribute table of the vector map was displayed in a raster map as a color-graded frequency distribution. The "Interpolation" plugin, which can also be used to interpolate the distribution between the recorded coordinates, was used in QGIS for this purpose.

3.8.3 Test drives Siebeldingen

During the design and development phase in 2013 and 2014, pure test drives were also carried out at the JKI in Siebeldingen. The main focus here was on checking the determination of stock coordinates, further processing, cooperation and compliance with the existing GPS system (Trimble) and the development and optimization of a suitable workflow. In principle, these trips were carried out in the same way as those in Geisenheim under the above-mentioned conditions and objectives.

3.8.4 Measuring trips Siebeldingen

After completion of the implementation phase (see section 4.3), an application trial was carried out in plot 23 with 2,726 vines in 43 rows. With a stand space of 2.10 m^2per vine, the area to be planted was around 5,725 m2 or 0.5725 ha. Based on the experience gained in image acquisition with the multi-camera system of the JKI, the trips took place at night (see Chapters 4.2.1 and 4.8), distributed over two consecutive dates. The logging functions of the control application were also activated during these measurements in order to record the absolute and relative accuracy parameters as well as the number of available satellites. To ensure the highest satellite availability, the vehicle's GPS antennas were `set to ant_up`.

3.9 Data evaluation and statistical analysis

The common spreadsheet programs Microsoft Excel 2010 under Microsoft Windows 7 and Libre Office 3 under Kubuntu 12.04.4 LTS, as well as the open-source statistical programming language R in version 3.1.2 were used for data evaluation and statistical analysis.

The results of the experimental tests from the measurement runs in Geisenheim and Siebeldingen were analyzed using various statistical methods corresponding to the measurement data.

To check the satellite availability, the distributions were compared with each other using the Wilcoxon test for pair differences.

Various descriptive statistical indicators were used to compare the RMS errors of the absolute and relative accuracies. The scatter and variance were assessed using the standard deviation sx and the coefficient of variation cv. The mean value comparison was carried out using the t-test for equal variances and the Welch test for different variances. The t-test requires normally distributed data, which is why the corresponding data sets were checked using the Kolmogorov-Smirnov test. An α error of at least 5% ($P = 0.05$) was defined as the significance level for all tests.

4 Results

This chapter first describes the results and the steps required to set up the final test carrier. Numerous tests were carried out during testing (see section 3.8), the progress of which is also briefly described. Finally, the results of the measurement runs in Geisenheim and at the JKI in Siebeldingen are presented.

4.1 Carrier vehicle

As part of the market evaluation before starting the design work, the available components for creating a suitable carrier platform were considered. To ensure a low-vibration drive, the choice of drive concept fell on electrically operated chain drives. Further exploration showed that an in-house design with its own drives, energy storage units and power and control electronics was not cost-efficient compared to a solution with a series vehicle.

4.1.1 Neo Alpine

At the same time as the market evaluation, the *Neo Alpin* large crate transporter (Fig. 30) from the South Tyrolean company Windegger was presented at Sitevi 2011, the trade fair for special crops held in Montpellier. This radio-controlled platform for transporting large crates during fruit harvesting is electrically driven. Wheel and track drives are available as alternative drive concepts. Two 12-volt lead-acid batteries are used as energy storage (system voltage 24 volts). The power and control electronics originate from the field of forklift trucks and are therefore robustly designed according to industrial standards. The vehicle width is 1 120 mm with a length of 1 140 mm. The manufacturer specifies a payload of 500 kg (WINDEGGER 2015) and a maximum driving speed of 3 km h^{-1}(WINDEGGER 2012). These dimensions already offer adaptation to the width of the alleys in vineyards. The low overall height and the installation of the accumulators close to the ground result in a low center of gravity.

In spring 2012, a Neo Alpin was purchased as the basis for the bonus system. After delivery, the maximum speed was measured on a test track of 30 m in the courtyard of the Institute of Technology. The result was an average value of 1.8 km h^{-1}.

Fig. 30 Large crate transporter Neo alpin from Windegger for transporting harvest crates (WINDEGGER 2015).

Initially, the plan was to realize a navigation solution with an RTK GPS system from Trimble in cooperation with Reichhardt Steuerungstechnik. For this purpose, the control board/microcontroller to be developed was to be connected to the encoder inputs of the Curtis impulse motor controllers installed on the Neo Alpin (Fig. 31).

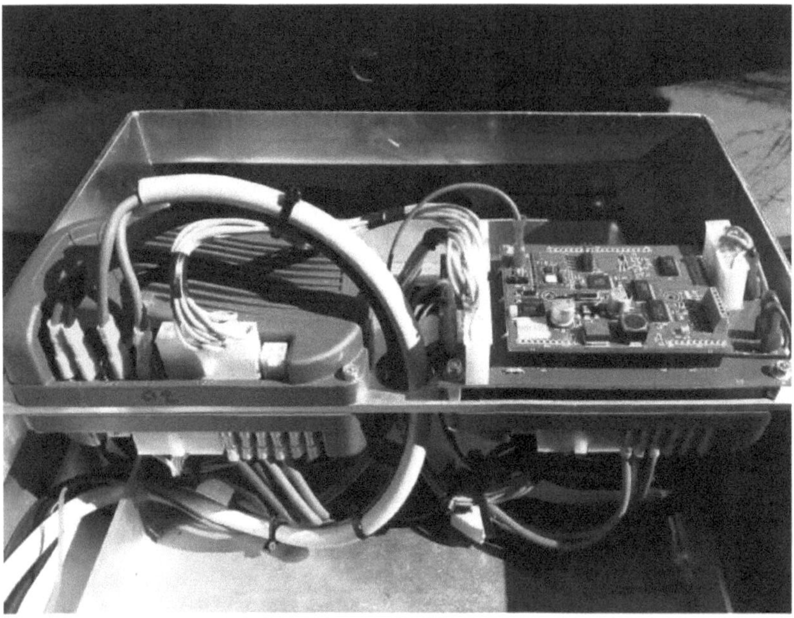

Fig. 31 Control and power electronics in the Neo Alpin.

4.1.2 Neosat Alpine

When contacting the Windegger company to clarify the technical details and wiring of the system, reference was made to the satellite-based navigation solution that was being developed at the time. In collaboration with an engineering office specializing in mechatronics (KMS), Windegger developed this for use in the Neo Alpin platform. This was followed by an agreement to upgrade the existing hardware and to jointly drive forward further development and adaptation to the application as a PHENObot.

In the course of the upgrade, the supply voltage of the vehicle was increased to 48 volts by installing two additional lead-acid batteries (Fig. 32).

Fig. 32 Batteries, power and control electronics (under the white cover) of the Neosat Alpin.

In addition, a 4-stroke internal combustion engine with 196 $cm^3$and 4.1 kW (HONDA 2015) was installed to drive a generator. This makes it possible to recharge the batteries during operation and extend the operating time. The generator can also be used reversibly to start the combustion engine electrically. On the one hand, manual starting is possible via the control program (see section 4.5) and, on the other hand, automatic starting is possible when the battery voltage falls below a defined value.

Two Hemisphere Eclipse II GNSS OEM modules with corresponding antennas were installed to realize the satellite-based navigation (Fig. 33).

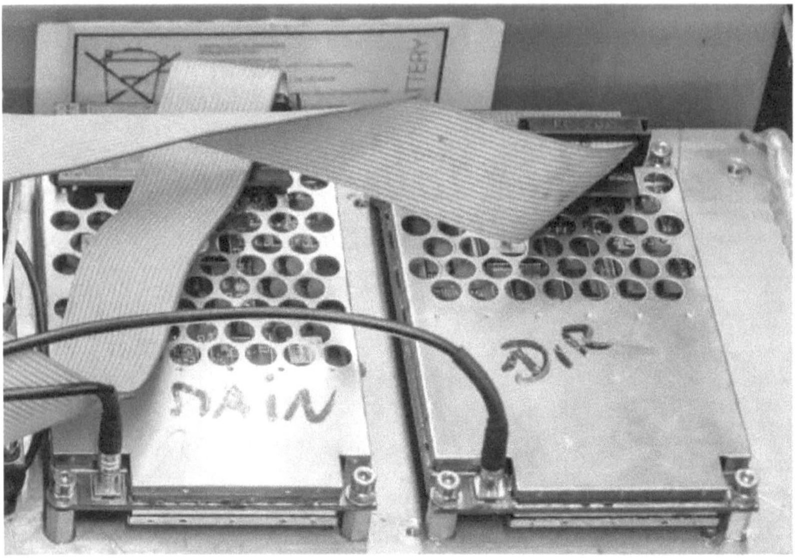

Fig. 33 Hemisphere Eclipse II GNSS OEM module. MAIN: Positioning. DIR: Direction determination.

In the current expansion stage (project completion 2014), only the NAVSTAR GPS system is used. The additional use of GLONASS would also be possible by purchasing the corresponding licenses. The receivers can use the L1 and L2 bands. The horizontal accuracy is specified by the manufacturer as 20 mm (HEMISPHERE 2012). The overall system is designed as RTK-GNSS and has a base station for correction signal transmission (Fig. 34). This also uses a module from Hemisphere as mentioned above. On the vehicle, one GNSS module is used to determine the position and the second to determine the direction. The second antenna is mounted horizontally at an angle of 90° to the direction of travel.

Fig. 34 Neosat Alpin during testing with RTK base station.

The vehicle is controlled via a microcontroller board developed by KMS (Fig. 35) (KOFLER 2015). The GPS modules, the radio antennas for receiving the correction signal and the remote control for manual operation, the impulse motor controls, the safety sensors and the drive motor sensors are connected to this board. A computer is connected via a USB interface for automatic driving operation.

The control system is divided into two parts. One part is responsible for integrating and controlling the hardware. Essentially, this includes

- Read/set the I/Os,
- Communication CanOpen motor driver,
- Communication/initialization GPS modules,
- Communication/initialization Radio modules,
- Control of drive motors,
- Regulation Generator,
- Control of manual operation,
- Safety functions (evaluation of ultrasonic sensors and mechanical safety shield).

All controllers and algorithms for automated driving are implemented in the second part of the control system. The *tasks* are realized with the help of so-called *tasks*:

- Task position controller,
- Task direction controller,
- Task position observer,
- Task direction observer,
- Task speed controller,
- Task Security,
- Task Communication.

During automated driving, the position controller guides the vehicle along a line that is defined by at least two given points (current position and at least one other point to be approached). A distinction is made between *linear* and *circular* trajectories. When approaching a point i, the vehicle first aligns itself along a circular trajectory (defined by a turning radius) to point i (direction controller). It then moves along the linear trajectory to point i. If several points are approached sequentially, there is always a change between circular and linear trajectories. Control takes place at 20 Hz. All points and lines of the travel tasks are stored in the controller in the form of a *route map*. The integrated route planner calculates the optimum route between the points to be approached.

KMS provided a simple test application and the necessary software libraries for the test operation. The software was developed for Linux-based operating systems. A Hewlett-Packard Compaq 6530b laptop with Kubuntu 12.04 LTS was used for the test drives in 2013. Due to the vehicle vibrations, this was equipped with a solid-state disk.

An inverter (48 V DC to 230 V AC) has been installed to supply the computer with power. The TeamViewer 8 or 9 remote maintenance software was used to remotely operate and monitor the control application while driving. Remote control was carried out via various Android end devices. A WLAN network with a router was used as the connection. To protect all external supply, control and connection components, a splash-proof and dust-proof switch cabinet was installed at the rear of the tracked vehicle (Fig. 34). At the end of the test operation without camera storage, an industrial PC (see chapter 4.3.2) was purchased and equipped with the necessary software (see chapter 4.4).

Fig. 35 Microcontroller board for control and parameterization.

4.2 Camera position control

To implement non-invasive scoring, the project partner JKI Siebeldingen set up and tested a multi-camera system (MKS) (Fig. 36). The first trials and reference scorings were carried out with a hand- or tractor-guided trolley (Fig. 37) and static mounting of the MKS.

The MKS consists of five cameras, three monochrome cameras, one RGB camera and one NIR camera (KICHERER et al. 2015). All cameras are installed in housings that are certified to IP67 and are therefore suitable for the requirements of outdoor applications. In the first construction stage, the MKS had a width of 600 mm, a height of 500 mm, a depth of 500 mm and a mass of 20 kg. The following requirements were defined for guiding the MKS on the PHENObot:

- Active positioning with four degrees of freedom: 2 translations vertically and horizontally; 2 rotations for tilting the MKS,
- Vertical positioning (measured to the center camera) of approx. 800-1 600 mm absolute to the floor,
- Horizontal distance of the focal plane to the foliage wall of 1 000 mm,

• Leveling (alignment in the vertical) before image acquisition Panning range per axis ± 15°.

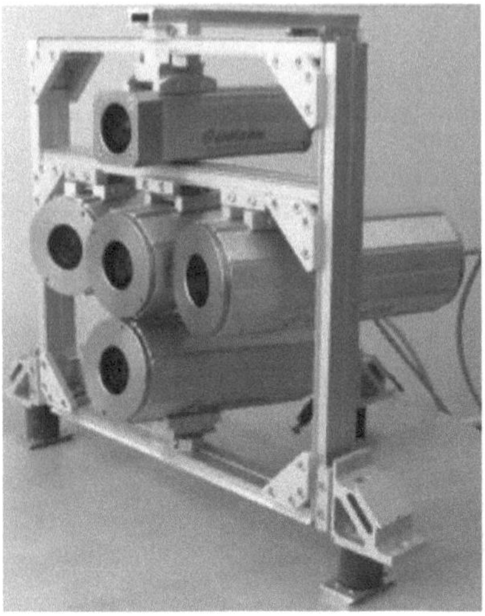

Fig. 36 MKS JKI Siebeldingen (KICHERER 2013b) Fig. 37 Hand cart for manual guidance of the MKS (KICHERER 2013b)

Based on these specifications, design drafts (3D) were created using SolidWorks 2011 CAD software.

4.2.1 Design I

The initial planning of the camera position control system was carried out to match the structure in Fig. 36, leaving sufficient installation space for a second MKS including camera guide. The design had to be correspondingly compact. Fig. 38 shows the first draft as a basic concept. The linear displacement paths along the horizontal and vertical were designed as spindle-operated linear displacements with cylindrical profiles. The rotational movements for levelling the MKS were to be reduced by gears. The design envisaged stepper motors as drive components for each axis. In the course of the initial planning, the concept was discussed with Nanotec, a supplier of linear and rotary drives, and suitable drive and control components were evaluated.

Fig. 38 Design I of the adaptive camera position control.

During the detailed planning phase, the first sampling results were evaluated with the MKS on the handcart. It became apparent that artificial lighting was necessary, as the changing position of the sun in particular had a considerable influence on the quality of exposure of the images taken (KICHERER 2013a). For this reason, the project partner JKI Siebeldingen purchased LED lighting bars and suitable mounting profiles for installation on the MKS (Fig. 39).

This considerably reduced the dimensions of the FMD with lighting (Fig. 40):

- Height: 730 mm,
- Width: 870 mm,
- Depth: 327.03 mm,
- Mass: approx. 60 kg.

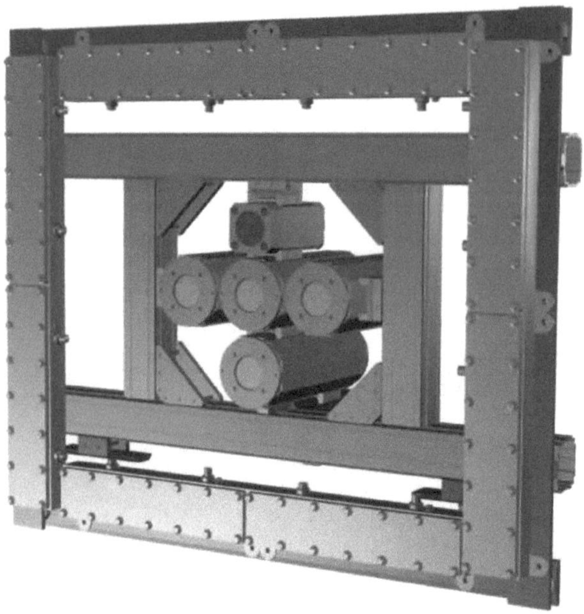

Fig. 39 Multi-camera system design II in 3D (AUTOVIMATION 2014).

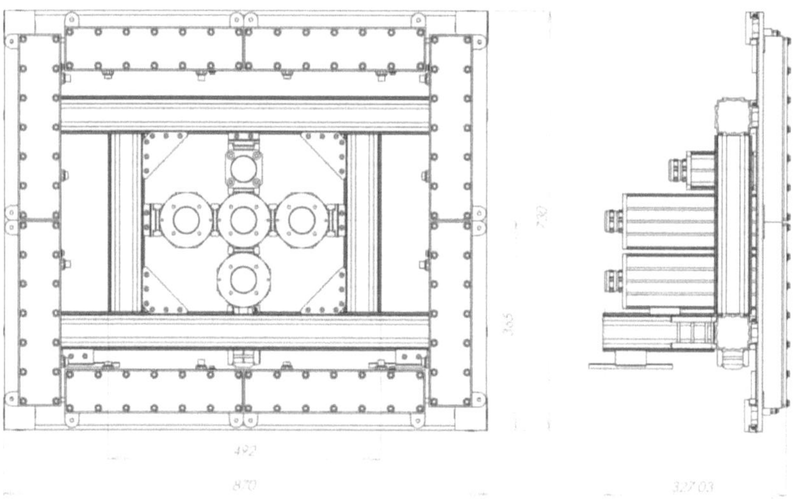

Fig. 40 Multi-camera system design II in 2D with dimensioning [mm] (AUTOVIMATION 2014).

4.2.2 Draft II

Due to the changed conditions, there was not enough space on the carrier platform to accommodate another FMD. For this reason, the concept was fundamentally changed during the creation of the second draft and the camera position control was positioned centrally on the vehicle. It was also decided to purchase all of the linear and rotary drives from third-party suppliers. Following a market evaluation, the design was commissioned from Igus in accordance with the specified framework conditions.

Fig. 41 Camera position control design II without drive motors.

Fig. 41 shows design II as a 3D model. This provided for laser-cut, folded aluminum sheets to accommodate the drive components.

(AlMg3 3.3535). For reasons of flexibility, this support concept was replaced by profile rails (Fig. 42) made of the aluminum alloy AlMgSi0.5 (3.32.06.72).

4.2.3 Draft III

The Institute of Technology had already worked with the BSO-GmbH profile system in the past. Therefore, tools and some leftover accessories and connecting material were available. For the construction of the base frame, the choice fell on the 40 mm grid with a groove diameter of 8 mm in a heavy design (PS40.4.0) and with a specific weight of 2.41 kg m^{-1}. The direct suspension of the MKS was realized according to the principle of *cardanic suspension* with profile rails of the PS40.12.0 system.

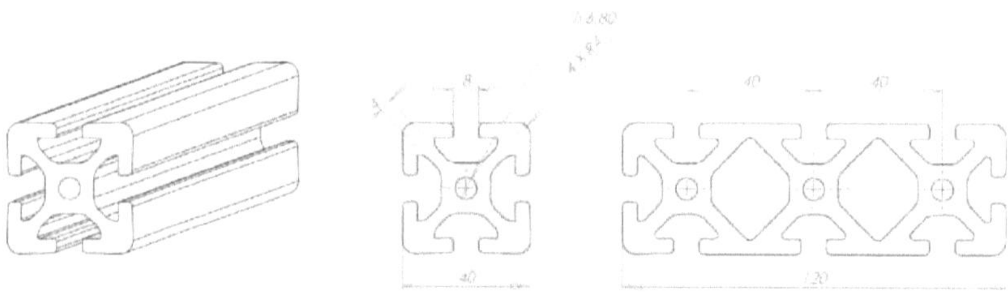

Fig. 42 Profile rails from BSO-GmbH; left and center: PS40.4.0; right: PS40.12.0;

Grid size: 40 mm; groove diameter: 8 mm; modified according to BSO-GMBH (2014).

Fig. 43 Design III with MKS; front view

Fig. 44 Design III with MKS; rear view

The originally selected supplier did not have any stepper motors in its product range to guarantee the forces required for the drive. To ensure long-term outdoor use, all drive components also had to be classified to at least IP67. After further market research, quotations were obtained for suitable stepper motors with IP68 classification on the coil side and on the drive shaft seal. The company Phytron offers suitable motors with the corresponding phyMOTION control series. These modular multi-axis motor controllers are equipped with their own main processors (CPU) and can accommodate the necessary digital and analog inputs and outputs as well as the required communication interfaces to higher-level systems (PHYTRON 2014). The proprietary phyLOGIC programming language is used for programming.

After reviewing the offers, all the required components were ordered from the companies BSO-GmbH, Igus and Phytron (Table 6).

Table 6 Mechanical and electrical elements (drive technology) of the camera position control (numbers in brackets = number of units).

Degree of freedom/element	***Mechanical element***	***Electrical element***
Horizontal (linear)	WSQ-20-1000 (2) WJ200QM-01-20 (2) WL200QM-01-20 LLZ (2) RGAS-JTRM-20x4 (1) SLS-20x4-FL (1)	ZSH 88/3.200.4-IP68 (1) INI-AB-I-025-B-AA (2)

Vertical (linear)	SLS-20x4-LL (1) PTGSG-20x4-01-R-1000 (1) 21 AT10/20-2/22H8 (2) TLK350 12x22 (2) AT10-600-16 (1) SLW-2080-AWM-800-ES (2) MF-3648-NEMA34 (2) COU-AR-K-120-120-32-32-B-AAA PRT-01-30-TO-AT10 (1) EFOM-20 (1)	ZSH 88/3.200.4-IP68 (2) IK-0009 (1)
Swivel 1 (rotary)	PN034-010 (1) 21 AT10/15-2/20H8 (1) TLK350 10x20 (1) AT10-560-16 (1)	ZSH 87/2.200.5-H 200-IP68 (1) INI-AB-I-025-B-AA (2)
Swivel 4 (rotary)	PRT-01-60-TO-AT10 (1) 21 AT10/15-2/20H8 (1) TLK350 10x20 (1) AT10-660-16 (1)	ZSH 87/2.200.5-H 200-IP68 (1) INI-AB-I-025-B-AA (2)
Control system	phyMOTION-10SL-W-s (1) Kübler 8.IS40.22321 (1)	POWM01.2 (1) MCM01.1 (1) I4XM01.1 (1) INAM01.1 (4) I1AM02.1 (1) DIOM01.1 (1) AIM01.1 (1)

The individual degrees of freedom are explained in more detail below. Components such as screws, pulleys and toothed belts are not included in the 3D models for reasons of scope and lack of CAD templates.

4.2.3.1 Horizontal

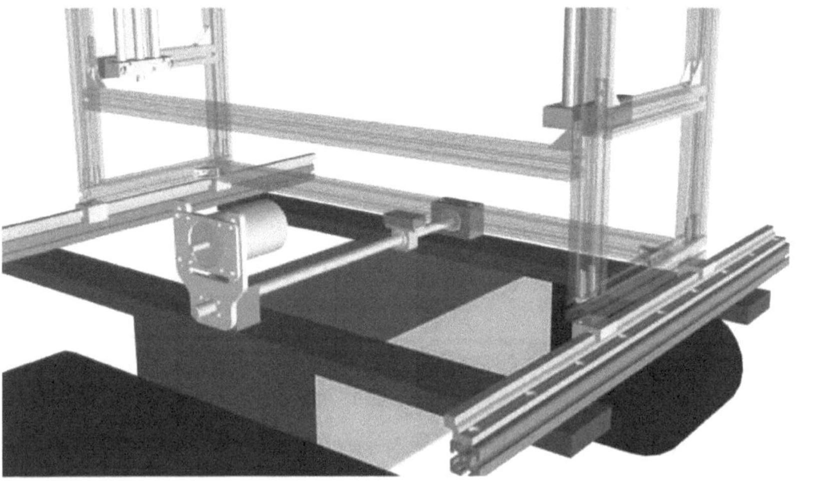

Fig. 45 Linear horizontal sliding device with guide rails (WSQ-20-1000) and stepper motor-driven spindle drive (PTGSG-20x4-01-R-1000).

To adjust the distance of the MKS in a horizontal direction to the foliage wall , a linear displacement in a dissolved design was provided. The entire base frame is slidably mounted on two rails made of hard-anodized aluminium with a square profile cross-section. Two fixed bearing blocks (WJ200QM-01-20) run on one rail and two floating bearing blocks (WL200QM-01-20 LLZ) on the other. According to the design, a drive force of at least 292 N is required. Using a threaded spindle (PTGSG-20x4-01-R-1000) with a pitch of 4 mm, a stroke length of approx. 500 mm and a feed rate of 0.5 m min^{-1}resulted in a required torque of 1.8 Nm. This torque is provided by a stepper motor (ZSH 88/3.200.4-IP68). For the supply

To reduce the installation space, the threaded spindle is driven by a toothed belt drive (21 AT10/20-2/22H8, TLK350 12x22, AT10-600-16) with a transmission ratio of i = 1:1, allowing the stepper motor to be rotated by 180° and thus not extend beyond the vehicle. Two adapter plates made of aluminum (AlZnMgCu1.5) were constructed to mount the stepper motor and adjust the belt tension (see electronic appendix). The detection of the end position and safety cut-off has been realized with two initiators (INI-AB-I-025-B-AA) at the outermost movement points.

4.2.3.2 Vertical

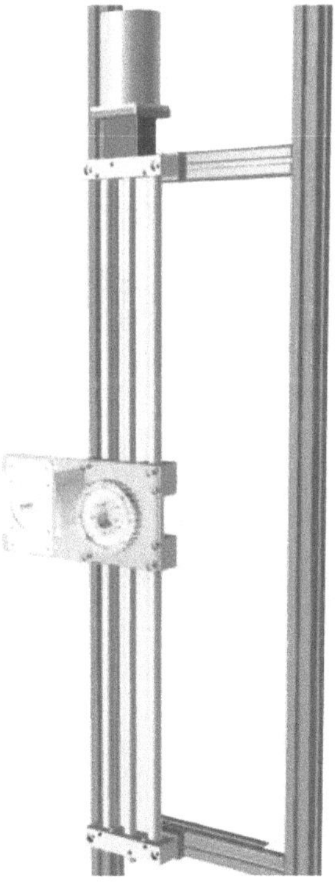

Fig. 46 Vertical sliding unit with linear tables (SLW-2080-AWM-800-ES).

The entire MKS can be moved vertically by 800 mm to adjust the image capture height. For this purpose, the concept provided for lateral mounting on two linear tables (SLW-2080-AWM-800-ES) made of hard-anodized aluminium with stainless steel spindles. These were attached to the sides of the base frame and are each driven by a stepper motor (ZSH 88/3.200.4-IP68). The weight force of the MKS plus the cardanic suspension (see 4.2.3.3) was assumed to be 800 N. According to the design, a minimum drive force of 441 N must be applied per linear table. A necessary drive torque of 2.46 Nm follows from the spindle parameters. The shafts were connected using claw couplings (COU-AR-K-120-120-32-32-B-AAA). To ensure the synchronization of the two linear units, the stepper motors were synchronized in the control system. A pre-assembled initiator /mounting adapter set (IK-0009) is used for limit switching.

4.2.3.3 Swivel axis I

The rotatable mounting of the MKS in a cardanic suspension (Fig. 47) serves to compensate for terrain inclination and unevenness. Both axes should each be able to swivel by ±15°.

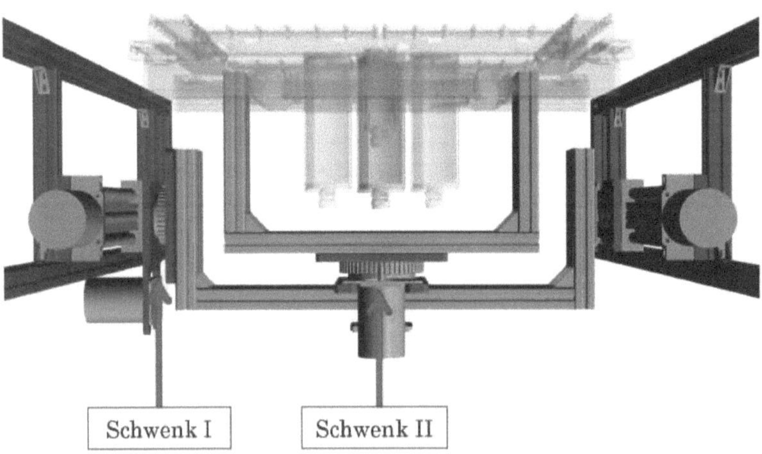

Fig. 47 Cardanic suspension of the MKS on the tables of the vertical linear units.

The pivot points of the first pivot axis parallel to the direction of travel (adjustment of the parallelism to the foliage wall) are located on the tables of the two vertical linear units. The bearing components are a simple plain bearing block (EFOM-20) on the passive side and a rotary table bearing (PRT-01-30-TO-AT10) with external toothing to accommodate the toothed belt (AT10-560-16) on the active driven side (Fig. 48). The geometry of the two-point suspension results in a central, radial bearing load of 400 N (assumed weight force of the MKS with cardan× 0.5). The stepper motor (ZSH 87/2.200.5-H 200-IP68) with flange-mounted planetary gearbox (PN034-010) (i = 1:15) was attached to the table of an SLW-2080-AWM-800-ES using two adapter plates. The swivel table is driven by a toothed belt drive with a transmission ratio of i = 1:2.27. This provides at least the torque of 31.2 Nm required for the drive. Detection of the end position and safety shutdown has been realized with two initiators (INI-AB-I-025-B-AA) at the outermost movement points.

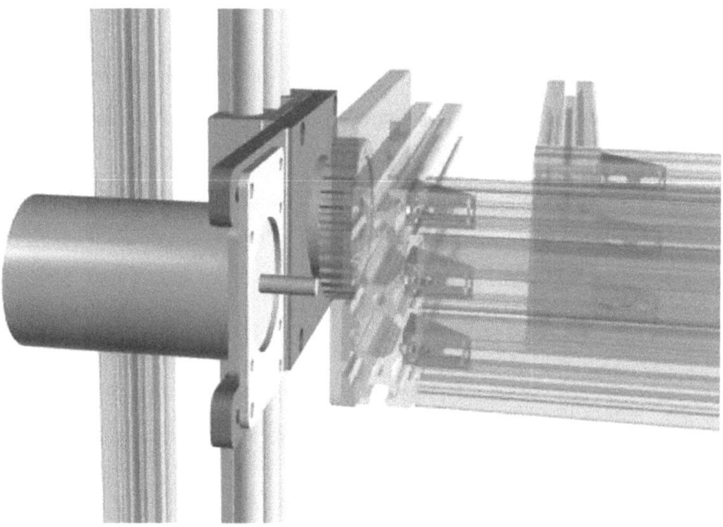

Fig. 48 Detailed view of swivel axis I with rotary table bearing, stepper motor, flange adapter and toothed belt pretensioning unit.

4.2.3.4 Swivel axis II

Fig. 49 Detailed view of swivel axis II with rotary table bearing, stepper motor, flange adapter and toothed belt pretensioning unit.

The second axis of rotation runs perpendicular to the direction of travel and can thus be used to compensate for the inclination in the fall or rise line. A rotary table bearing of a larger design (PRT-01-60-TO-AT10) was provided for this purpose (Fig. 49). It is driven by a stepper motor (ZSH 87/2.200.5-H 200- IP68) via a toothed belt drive (21 AT10/15-2/20H8, TLK350 10x20, AT10-660-16). Two initiators (INI-AB-I-025-B-AA) were again required for limit switching.

4.3 PHENObot

Once the technical design and planning had been completed, the scoring system was set up. Fig. 50 shows a three-page view of the vehicle dimensions according to the concept.

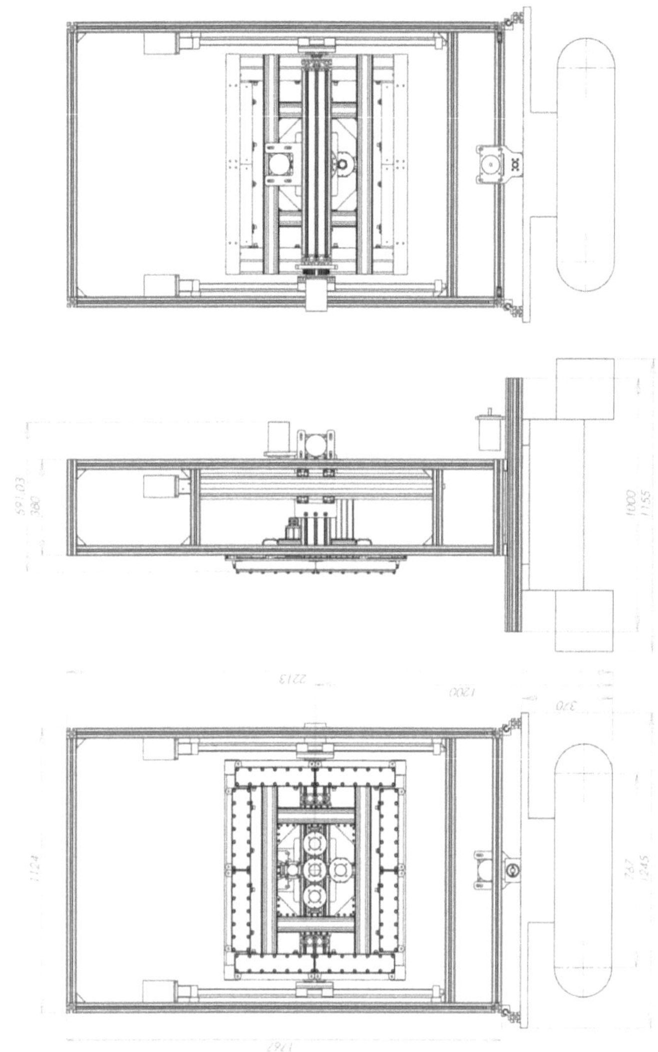

Fig. 50 Front, side and rear view of the PHENObot with dimensioning [mm] to show the dimensions.

4.3.1 Assembly and installation

Once all the components had been delivered, the camera control system was installed and the electrical components were wired. Deviating from the design draft, some changes had to be made during assembly due to the structural conditions on the vehicle, especially on the box slide, and due to the generator unit.

First, the base frame (Fig. 51) was made and placed on the rails for horizontal displacement. For better fastening, the rails with the support profiles were turned 180° inwards. In order to be able to remove the entire structure with MKS for maintenance work, for example, profiles were inserted for the cross connection (Fig. 52).

Fig. 51 Mounting the base frame of the camera position control. Horizontal displacement.

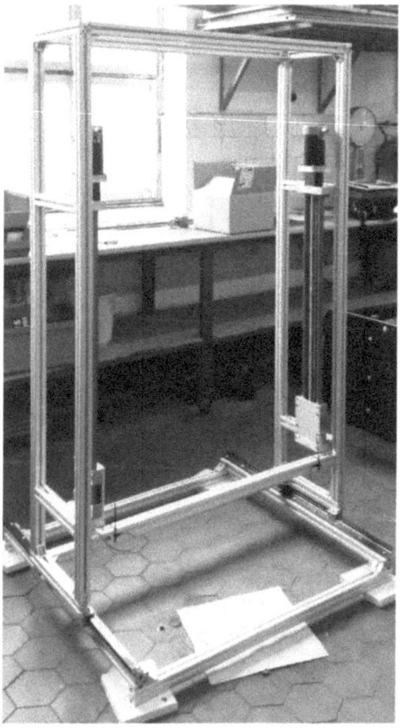

Fig. 52 Assembly of the base frame of the cam position control. Vertical displacement.

The cross connectors were extended to accommodate the control unit with switch box and control panel (Fig. 53).

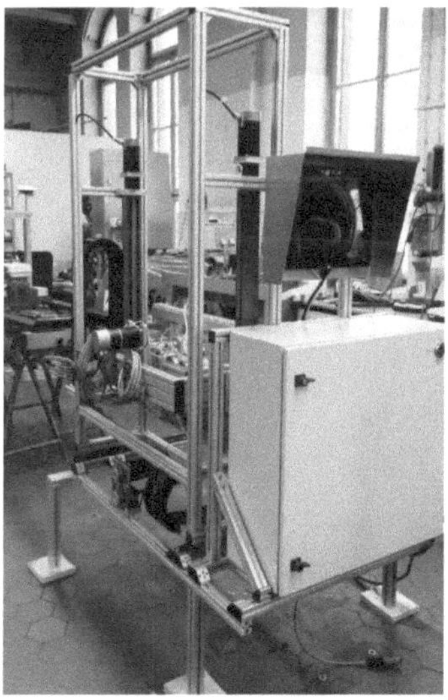

Fig. 53 Camera position control assembly as a remote unit. Rear with operating unit.

Fig. 54 Camera bearing control assembly as a remote unit. Front with MKS.

The remote camera bearing control system can be installed with four supports. The entire vehicle is lifted with a fork lift truck for disassembly. Two screws are loosened at each of the four fixing points (Fig. 55). After disconnecting the cable connections (see chapter 4.3.2), the vehicle is lowered down to the seat base, the chassis is still suspended and can be removed completely with the lift truck.

Fig. 55 One of the four attachment points of the camera position control system on the vehicle's support frame.

Energy chains are used to guide the control and power cables along the *horizontal* and *vertical* adjustment axes (Fig. 56, Fig. 57 and Fig. 58).

Fig. 56 Cable routing horizontal displacement.

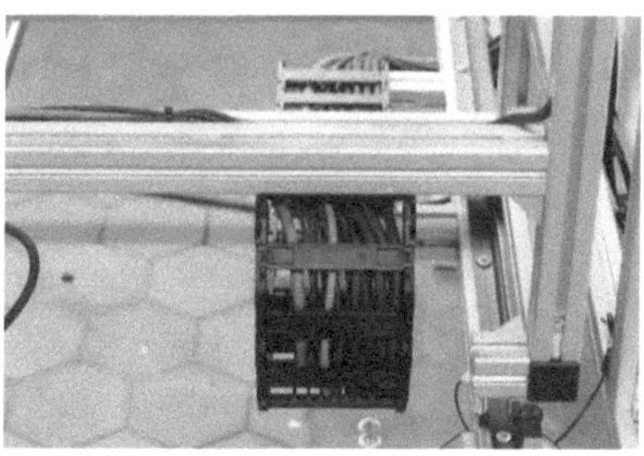

Fig. 57 Cable routing horizontal displacement.

Fig. 58 Cable routing vertical displacement.

4.3.2 Electronics

The electrical and electronic parts and components on the PHENObot are divided into supply and control.

4.3.2.1 Supply

A 48 V DC system is available on the carrier vehicle as an on-board power supply. Four lead-acid batteries connected in series, each with 12 V and 55 Ah, serve as the voltage source (Fig. 32). The battery pack is connected to the vehicle system via an Anderson SB50 high-current plug connection. The manufacturer requires this plug connection to be disconnected for charging. The batteries can then be connected to the charger supplied. If the camera position control system is fitted, this connection is not accessible. For this reason, an additional connector plug for charging the batteries has been installed in a splash-proof housing (Fig. 59). Access to the plug is secured with a lock. To be able to connect the charger, the ignition key (Fig. 60) of the vehicle must be removed and access opened with it. This ensures that the vehicle electronics do not inadvertently remain switched on during the charging process. Otherwise damage could be expected.

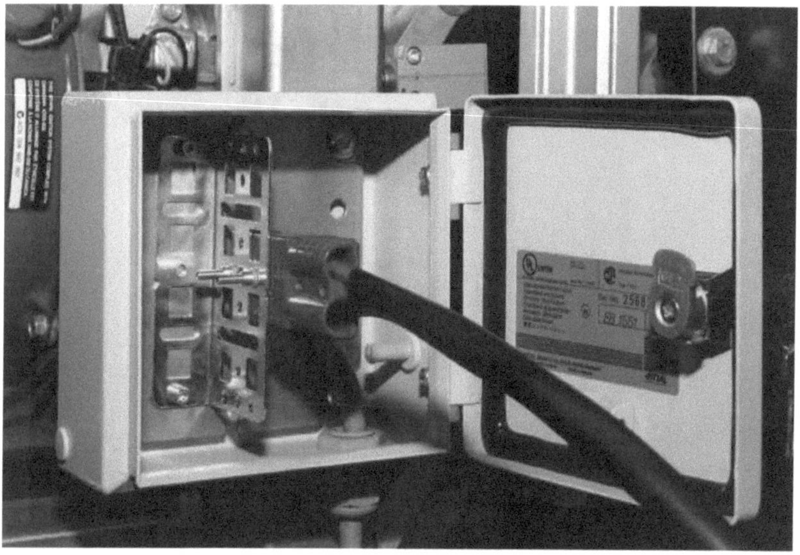

Fig. 59 Anderson SB50 connector plug with charging cable for charging the batteries.

An 80 A back-up fuse was installed as a protective element between the battery pack and the power electronics for the traction drive. The power supply for the switch cabinet in the rear area (camera position control, navigation and image computer) was protected with a 30 A fuse. The connection to the vehicle can be disconnected via a plug connection. To minimize the switching current, the microcontroller is supplied with the system voltage when the ignition is switched on. The firmware boot process is initiated. At the end of this process (after approx. 10 seconds), a control voltage is applied to a relay that switches on the load circuit with 48 V. The vehicle is then ready to drive in manual mode via remote control.

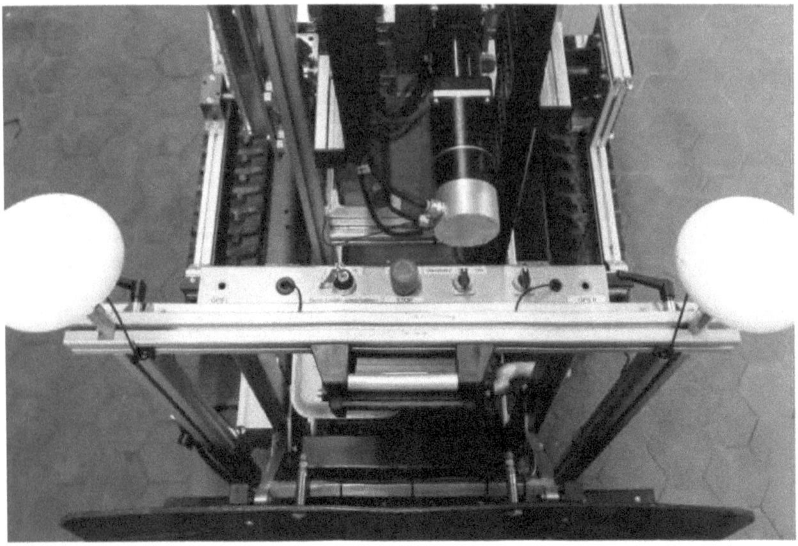

Fig. 60 Vehicle control unit, from left: GPS 1, ignition lock, emergency stop, generator free, not used, GPS 2.

4.3.2.2 System layout PHENObot

Fig. 61 shows the entire structure of the electrical/electronic system of the scoring system. The PHENObot basically consists of:

- Carrier vehicle,
- Navigation system with computer I (interface to the vehicle),
- Camera position control,
- Camera system with computer II (interface to the navigation system),
- Control unit (control cabinet with control panel and electrical supply).

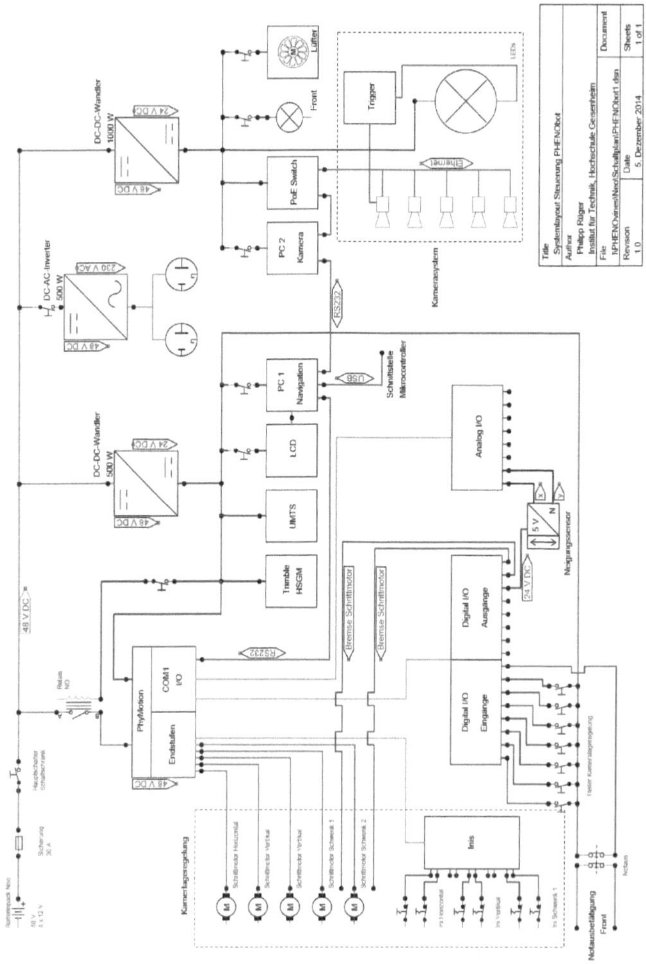

Fig. 61 System layout controlPHENObot.

The carrier vehicle and its components have already been described in detail in section 4.1. The navigation algorithms are executed on the vehicle's microcontroller. A computer (PC 1) is required for target specification and operation. A spo-book MOVE T56N resistive industrial PC (IPC) from spo-comm with x86 architecture, 8 GB RAM, a 120 GB solid state disk, WLAN and UMTS modules and an 800 mAh battery pack was purchased for this purpose (PC 1). The housing complies with the MIL-STD-810F standard. The connection to the microcontroller is established via the USB interface . Further connections exist via the RS232 serial interfaces to the stepper motor controller and to the image computer (PC 2). This computer was provided by the JKI, which is why the specifications will not be discussed in detail here.

Fig. 62 Front panel of the phyMOTION stepper motor control unit.

The camera position control consists of the electrical movement components (stepper motors), the controller (phyMOTION, actuation switch) and various sensors (initiators, tilt sensor). Five hybrid stepper motors, each with two phases in bipolar operating mode, were installed. The two stepper motors of the swivel axes also have motor brakes and incremental encoders. On the load current side, the motors are connected to the stepper motor output stages of the phyMOTION (Fig. 62) with five-core cables. Two conductors with a cross-section of 1 $mm^2$are available for each phase. The windings in the stepper motors are connected in parallel. The fifth wire is the protective conductor. The motor brakes are connected to the digital output module of the phyMOTION using two wires with a cross-section of 0.5 mm2 . Each movement axis has a pair of initiators for end position switching. These were connected to the corresponding inputs of the phyMOTION using three-wire initiator connection cables with a cross-section of 0.25 mm2 . A Kübler 8.IS40.22321 biaxial inclination sensor (Fig. 63) is used to detect the position of the MKS. This was connected with four wires with a cross-section of 0.5 mm2 to two analog inputs of the phyMOTION.

Fig. 63 Inclination sensor Kübler 8.IS40.22321.

The multi-camera system is connected to a PoE network switch (Fig. 64) with CAT6 Ethernet cables and supplied with power via this. The eight LED lighting bars are connected in series in two circuits of four units each and are triggered simultaneously by one of the five cameras via a trigger box (Fig. 64). The camera switch and trigger box are each supplied with power using a 2×1.5 mm^2cross-section.

Fig. 64 Housing network switch MKS (left) and housing (right) Trigger LED lighting with power supply (center).

Three voltage ranges are provided in the control unit (Fig. 65):

48 V direct current (directly from the vehicle)

- Stepper motors via stepper motor output stages,
- Voltage converter for generating the other voltage ranges (DC-DC or DC-AC converter).

24 V DC voltage (circuit I; DC-DC converter 500 W)

- Navigation components,
- Control panel,
- WLAN router,
- Secondary GPS,
- I/O supply phyMOTION.

24 V DC voltage (circuit II; DC-DC converter 1000 W)

- MKS with PC 2,
- LED lighting,

- Headlights (LED),
- Enclosure fan.

Fig. 65 Interior view of the control cabinet with all components and wiring.

4.3.2.3 Control system

There are two operating areas on the vehicle to control the bonus system. The functions of the carrier vehicle are selected in the front area (Fig. 60). In the rear area, a switch panel (Fig. 67 and Fig. 68) and a capacitive touch-sensitive LCD screen (Fig. 66) for operating the two computers are located on the control cabinet (Fig. 65).

Fig. 66 Control panel control cabinet with touch LCD and mouse/keyboard combination. The rover of the RTK system from Trimble with TSC 1 is also installed here.

The touch LCD is used to control all relevant program functions of PHENObotControl 1.0 (see chapter 4.4) and IGG geotaggers on the image computer. In addition, a mouse/keyboard combination is available for special functions and extensive text input. The operating and display elements are connected to both computers via a switch (KVM switch).

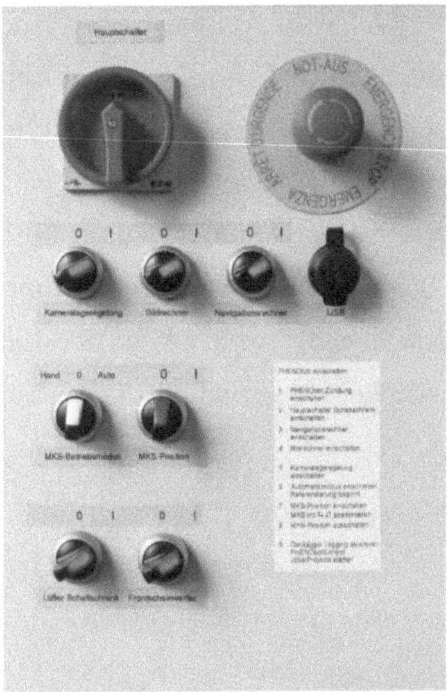

Fig. 67 Switch panel I

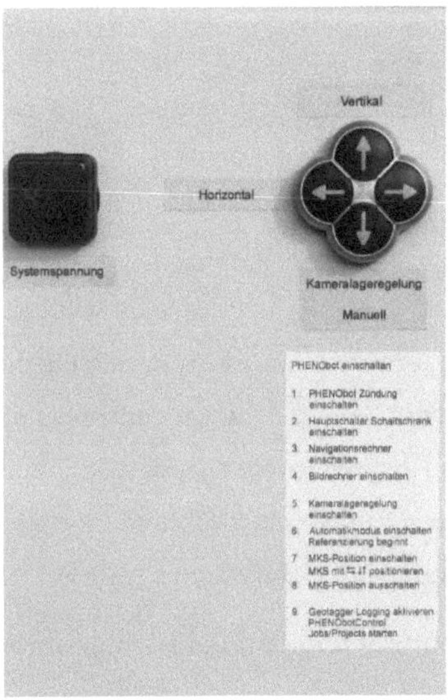

Fig. 68 Switch panel II

The control cabinet is supplied with the system voltage via a main switch. If this is switched on, the two DC-DC converters start and provide 24 V DC voltage. All other hardware components can be switched on separately via rotary switches (Fig. 67). The respective supply voltage is applied to the switches; closing the switches supplies the components at the switching inputs with voltage. With the phyMOTION stepper motor controller, a relay is connected in between, as the current rating exceeds the specification of the switches used. The PC 1 navigation computer has a battery pack to compensate for possible voltage interruptions. When switched off using the corresponding rotary switch on the control panel, the computer continues to run on battery power for five minutes. The computer then shuts down automatically. The DC-AC converter has its own switch located directly on the device, which is why an external switching option on the control panel has been omitted here.

The two switches *MKS operating mode* and *MKS position* (Fig. 67), as well as the control cross (Fig. 68) are connected to digital inputs (EX.X) of the phyMO- TION. If these are switched on, 24 V is applied to the respective input, so the control unit evaluates this as set (value = S; unset = R).

- Manual FMD operating mode: E1.7 = S; E1.8 = R
- MKS operating mode Auto: E1.7 = R; E1.8 = S
- MKS operating mode 0: E1.7 = R; E1.8 = R
- MKS position 0: E1.2 = R
- MKS position 1: E1.2 = S

The emergency stop switches at the front and rear do not have a voltage-disconnecting function. Only any movement (camera position control and vehicle movement) is interrupted. When the system is switched on again, it continues at the interrupted point. The two emergency stop switches both have the same function. The front switch is connected to the control cabinet via a cable connection. This can be disconnected via a plug connection (removal of the camera frame).

The voltage indicator in the control cabinet is connected in parallel to the vehicle accumulators and thus indicates the system voltage.

4.3.3 Stepper motor controller phyMOTION

The stepper motors on the movement axes are controlled by the phyMOTION stepper motor controller (Fig. 62) to control the position of the MKS. This has five output stages as well as integrated indexers for controlling and synchronizing the two stepper motors for the vertical movement of the MKS. The stepper motor controller contains its own CPU and can be connected to a PC via a USB port. The phyLOGIC Toolbox configuration program can be used to parameterize the controller (Fig. 69). With phyLOGIC, Phytron provides its own programming language for creating complex motion tasks. The corresponding programs are interpreted and executed by the phyMOTION. The `PHENObot_auto.mpr` program was written in phy LOGIC and loaded into the phyMOTION's memory for levelling the FMD and for manual operation using the control pad (Fig. 68). This is executed automatically each time the control unit is started. In the main loop, the position of the *MKS operating mode* rotary switch is queried (Fig. 70). In *Manual* mode, the horizontal and vertical axes can be set using the control pad. In *Auto* mode, the swivel axis I starts referencing to the initiator M0P+, sets the position counter at this limit switch to zero and moves to the center position (horizontal). The *horizontal* and *vertical* axes then start referencing in parallel. The *horizontal* axis moves the FMD to the greatest possible distance from the potential foliage wall, vertically it is moved to the lowest position. The MKS can then be moved manually to the desired height and distance from the foliage wall. To do this, the *MKS position* switch is switched on and the control pad is released. Once positioning is complete and the *MKS position* is switched off again, leveling commands can be received from the PHENObotControl application on the navigation computer via the serial interface. This RS232 telegram sets register R99 in the phyMOTION to the value 1. This starts the automatic leveling in triple repetition according to the measured values of the inclination sensor Kübler 8.IS40.22321 (Fig. 63). After levelling is complete, the position of the *FMD operating mode* switch is queried again and the corresponding program loop is called up.

Fig. 69 phyLOGIC Toolbox with excerpt from the PHENObot_auto.mpr program for controlling the levelling.

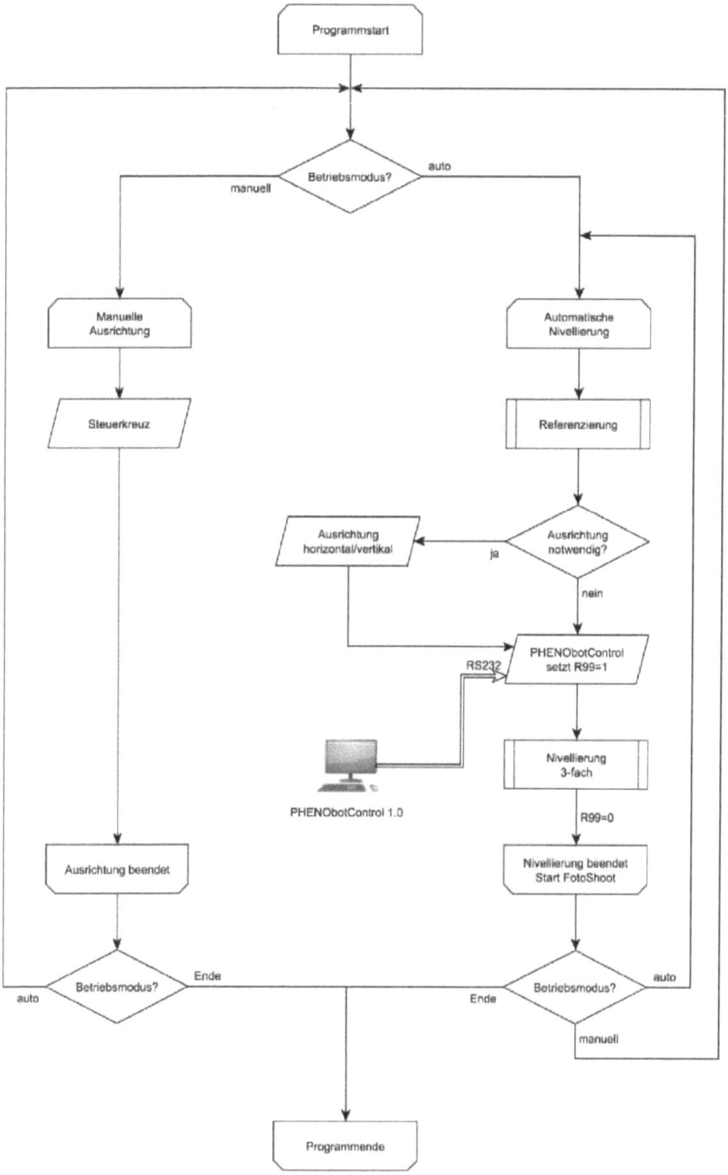

Fig. 70 Program sequence stepper motor control PHENObot_auto.mpr.

4.4 Implementation

Once the assembly work on the camera position control system had been completed, the FMD was implemented in the carrier vehicle. For this purpose, the profile frame of the lighting unit was firmly bolted to the inner rotating fork of the cardanic suspension. The final step was the production of the power supply and the data connection. The description of the detailed test phase follows in chapter 4.8.

4.5 PHENObotControl

Based on the experience gained with the test application from KMS, we developed our own control application (Fig. 71). The C++ application uses the library provided by KMS to process and control parameters and tasks.

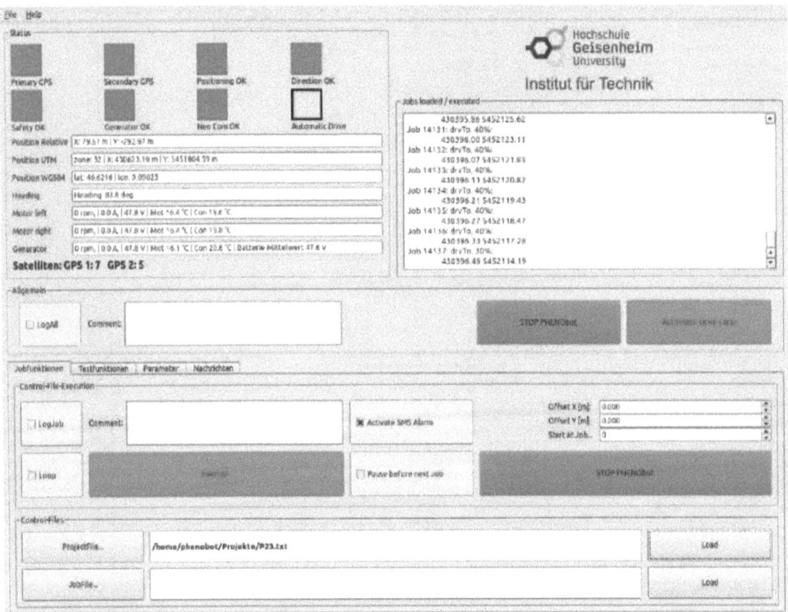

Fig. 71 PHENObotControl 1.0 control application (final release).

The movement and parameter control was defined by KMS in the library `libneosat.so` and the associated header files `neosaterror.h`, `neosatjob.h` and `neosatlib.h`. Fig. 72 provides an overview of the defined structures, data types and functions of the entire application. The green and gray fields contain the type definitions, the structures are described in the other fields, the functions of the actual application in the yellow field. The Qt library is integrated as a C++ class library to implement the graphical user interface. The *widgets* module in particular was used to create the GUI. The elements of the GUI are linked to the program functions according to the *signal-slot concept* of Qt. The following source code files were generated:

- `main.cpp/.h`,
- `mainwidget.cpp/.h`,
- `rs232neo.cpp/.h`.

The main program is executed in the `main.cpp` file. This is defined in the `mainwidget.cpp` file and is called as the `MainWidget()` function in the `main.cpp` file.

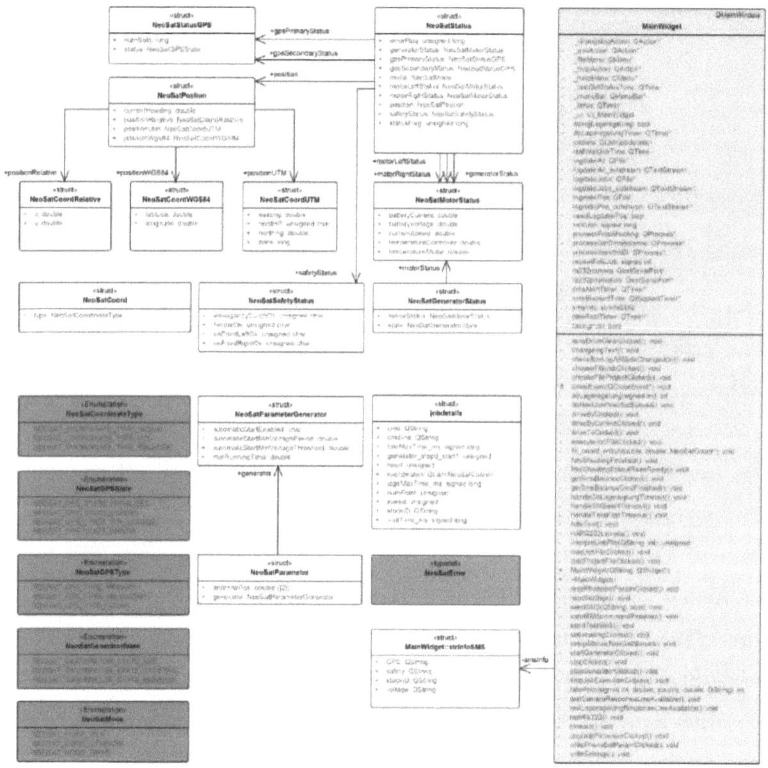

Fig. 72 Overview of the program functions, structures and data types of the application PHENObotControl.

The central application of the control program is the interpretation of travel commands and other tasks such as initiating the MKS level control and triggering the MKS. The QextSerialPort library was implemented for communication with the phyMOTION stepper motor controller and the image computer via two serial interfaces. The position control and image draw functions are defined in the `rs232neo.cpp` file. The function call is made from the file

`mainwidget.cpp`.

The execution of the orders is initiated by commands. These must be available together with coordinates in a text file (see chapter 4.6). Possible commands are

- `drvTo,[...]` (travel order with speed specification [%]),
- `setHd,[...]` (setting a heading with specification of an angle),
- `wait,[.]` (wait time with time specification [ms]),
- `doCmd,[.]` (execution of a command line command),
- `foto,[.]` (triggering the image computer with a waiting time [ms]),
- `position,[.]` (triggering of position control with waiting time [ms]),
- `logPos` (to create a log file, see section 3.7.1),
- `generator,[.]` (switch on [Arg=1] or switch off [Arg=0] generator).

The text files (job lists) are created with six columns (Fig. 73). The first two contain the UTM coordinates (X, Y), which are interpreted by the `drvTo command`. Columns three and five contain indices for sorting the points when exporting from the GIS software (see Chapter 4.6). An identification option (`stockID`) can be inserted in column four. The last column contains the commands. All commands are interpreted and processed sequentially. In the example (Fig. 73), the coordinate of the floor with the ID 23-001-072 is approached in line 2 at a speed of 30% of the maximum vehicle speed. This is followed by a waiting time of 200 ms, after which the leveling of the FMD is triggered. After the waiting time of 4,500 ms, the image computer receives the telegram for triggering the FMD. The success message for the image draw must be sent within 6,000 ms. Finally, the next job is executed after a waiting time of 500 ms.

```
1  430476.169500|5452118.773100|73|23-001-E|72|drvTo,30,wait,500
2  430475.996800|5452122.068600|72|23-001-072|71|drvTo,30,wait,200,lage,4500,foto,6000,wait,500
3  430475.938458|5452123.140772|71|23-001-071|70|drvTo,30,wait,200,lage,4500,foto,6000,wait,500
4  430475.866781|5452124.458021|70|23-001-070|69|drvTo,30,wait,200,lage,4500,foto,6000,wait,500
5  430475.806218|5452125.571011|69|23-001-069|68|drvTo,30,wait,200,lage,4500,foto,6000,wait,500
6  430475.738166|5452126.821648|68|23-001-068|67|drvTo,30,wait,200,lage,4500,foto,6000,wait,500
7  430475.677321|5452127.939812|67|23-001-067|66|drvTo,30,wait,200,lage,4500,foto,6000,wait,500
8  430475.551140|5452130.272039|66|23-001-066|65|drvTo,30,wait,200,lage,4500,foto,6000,wait,500
9  430475.508391|5452131.044333|65|23-001-065|64|drvTo,30,wait,200,lage,4500,foto,6000,wait,500
10 430475.444686|5452132.215057|64|23-001-064|63|drvTo,30,wait,200,lage,4500,foto,6000,wait,500
11 430475.378618|5452133.429237|63|23-001-063|62|drvTo,30,wait,200,lage,4500,foto,6000,wait,500
12 430475.310855|5452134.674542|62|23-001-062|61|drvTo,30,wait,200,lage,4500,foto,6000,wait,500
13 430475.243134|5452135.919081|61|23-001-061|60|drvTo,30,wait,200,lage,4500,foto,6000,wait,500
14 430475.125461|5452138.095084|60|23-001-060|59|drvTo,30,wait,200,lage,4500,foto,6000,wait,500
15 430475.083049|5452138.861043|59|23-001-059|58|drvTo,30,wait,200,lage,4500,foto,6000,wait,500
16 430475.024561|5452139.935907|58|23-001-058|57|drvTo,30,wait,200,lage,4500,foto,6000,wait,500
17 430474.954593|5452141.221746|57|23-001-057|56|drvTo,30,wait,200,lage,4500,foto,6000,wait,500
18 430474.893575|5452142.343092|56|23-001-056|55|drvTo,30,wait,200,lage,4500,foto,6000,wait,500
19 430474.827284|5452143.561369|55|23-001-055|54|drvTo,30,wait,200,lage,4500,foto,6000,wait,500
20 430474.707915|5452145.768544|54|23-001-054|53|drvTo,30,wait,200,lage,4500,foto,6000,wait,500
```

Fig. 73 Section of a job list (Geilweilerhof, plot 23, row 1) with coordinates, sequence indices, stockIDs and commands.

In this case (Fig. 73, line 2), the telegram that is sent to the image computer via the serial interface is composed as follows:

`$PosReached, 430475.9968  5452122.0686 357.1 23-001-072 '.`

The keyword `$PosReached` is used to signal to the IGG Geotagger image acquisition program on the image computer that the position for image acquisition in front of floor 72 in the first row of plot 23 has been reached and that the image can be drawn. The coordinates used here for X and Y were taken from the job file (Fig. 73) as an example. In the real case, these are taken from the GPS antenna I that provides the position and must be corrected by the vehicle orientation (357.1°) and the offset parameters in order to obtain the position in relation to the center of the vehicle (see chapter 3.8). If the image acquisition was successful, the image computer sends another telegram back to the navigation computer via the serial interface with the content `$OK`. This message must be received within the timeout threshold defined in the job file with the photo command (e.g. 6,000 ms). Otherwise, the application assumes an error during image capture and sends the photo command again. If there is no positive response from the image computer in this case either, further job execution is paused and an SMS alarm is sent. The user must rectify the error during image capture and end the pause. Another attempt is then made to trigger image acquisition at the same position. This procedure is repeated until a positive response is received. The IGG Geotagger recording software stores the data from the telegram together with position data from the Trimble RTK system in a text file in the image folder. This allows the position data and the ID to be clearly assigned to the images during image evaluation.

Despite object-oriented programming, a program flow diagram is shown in Fig. 74 to make it easier to understand how the control application works.

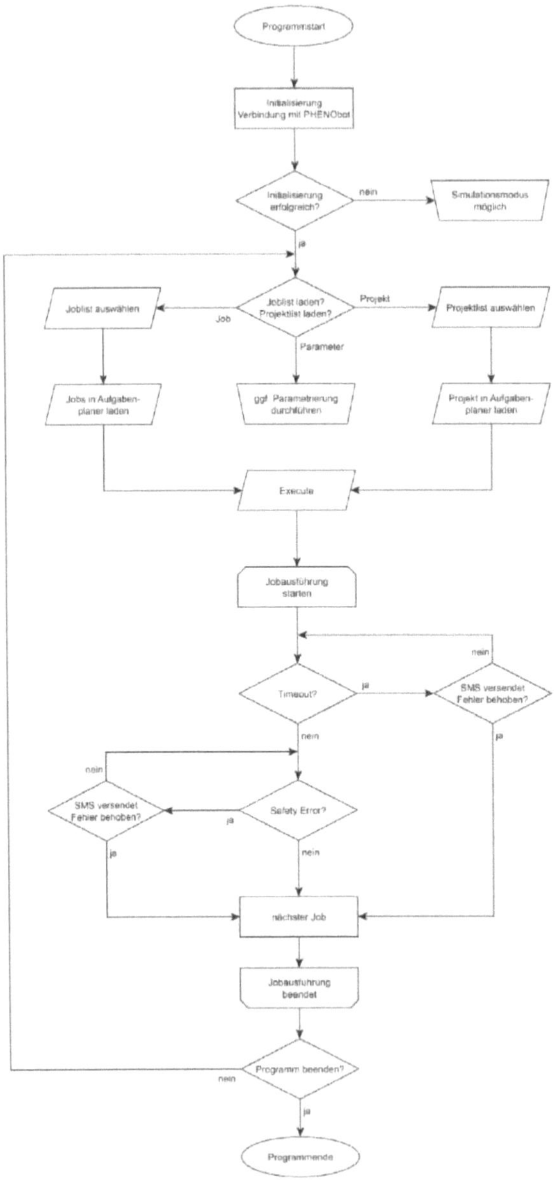

Fig. 74 Program sequence PHENObotControl 1.0

After starting the program, the connection to the microcontroller and the vehicle status are checked (Fig. 75).

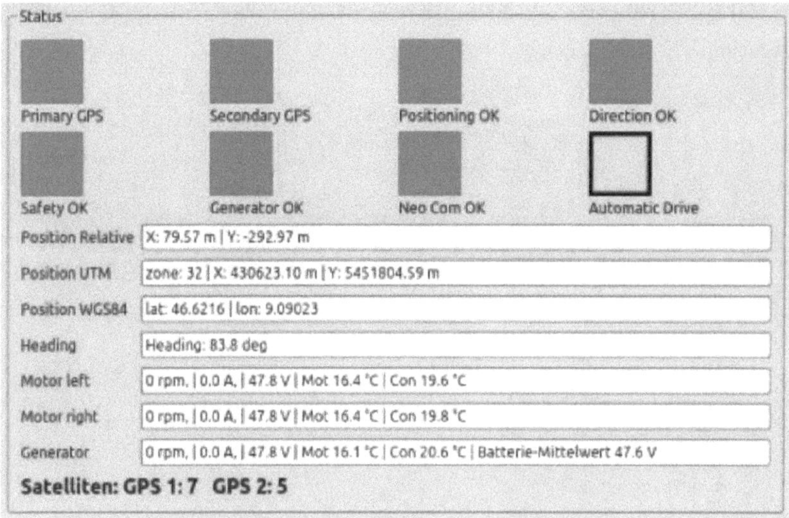

Fig. 75 Status area PHENObotControl 1.0 after successful initialization.

If the result is negative, simulation mode is still possible. For this purpose, a checkbox to ignore the vehicle status can be selected in the program. If initialization is successful, job or project lists (see chapter 4.6) can be selected and loaded (Fig. 76).

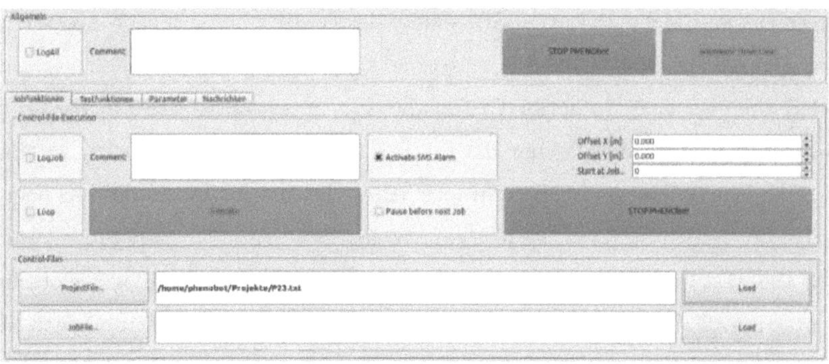

Fig. 76 Central control panel with selection area of the job or project list in the lower area.

These are loaded into the application's task planner by pressing the *Load* button (Fig. 77). The log functions `logAll` and `logJob` as well as the repeat function

`Loop` can be activated by selecting checkboxes and SMS alerting can be deactivated if desired. A comment can be entered for the log functions, which is inserted into the file name of the log file and the header. Pressing the *Execute* button starts the job execution with acknowledgement. The *Stop* button cancels the execution.

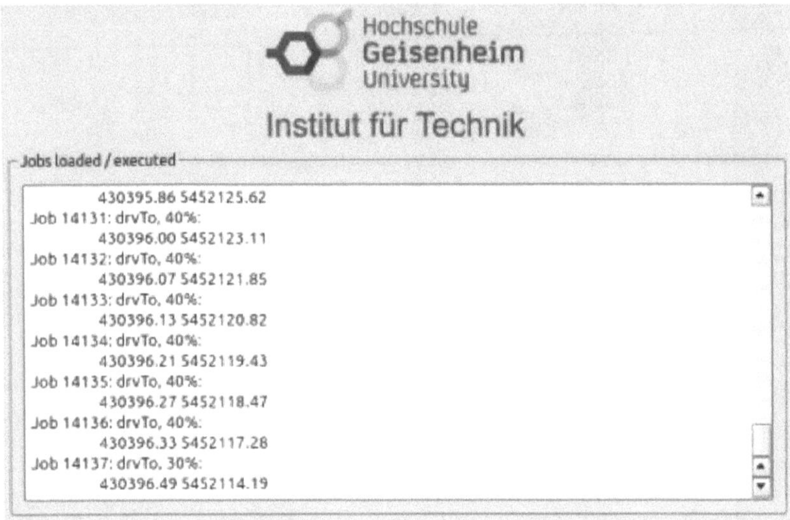

Fig. 77 Task planner, during execution the sequential processing is displayed step by step.

If SMS alerting is activated, a countdown timer starts with each individual job until the job must be completed. If this is not the case, execution pauses and the user receives an SMS notification. If a safety device (ultrasonic sensor, approach sign or emergency stop switch) is activated and the timeout threshold is also exceeded, the user is again notified. After the error has been rectified by the user and acknowledged by deselecting the pause function (automatic setting in the event of an error), job execution continues. After the last job in the list has been executed, the vehicle stops at the last defined coordinate and is ready to load and execute the next list. The termination of the job list also closes the log file of the `logJob function`. The loop function enables the endless execution of a job or project list (e.g. for presentation). The program interface can be closed by the user once all tasks have been completed.

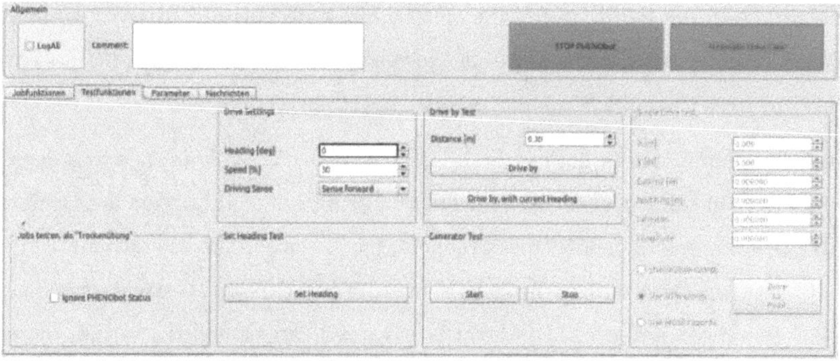

Fig. 78 Test functions tab of PHENObotControl 1.0

In the *Test functions* tab (Fig. 78) of the application, you can try out various guessing functions. By activating the *Ignore PHENObot Status* checkbox, the simulation mode can be selected without a connected PHENObot or without sufficient satellite reception in order to check job or project files for consistency in advance. In the *Drive Settings* area, the vehicle can be manually brought into a desired orientation at a given turning speed. The *Drive by Test* mode enables manual driving around a specific route either with the parameters from the *Drive Settings* or with the current vehicle orientation in decimal degrees. In addition, the generator can be started or stopped manually. The *Single Drive Test* area was only required during the test phase and has therefore been deactivated in the final version.

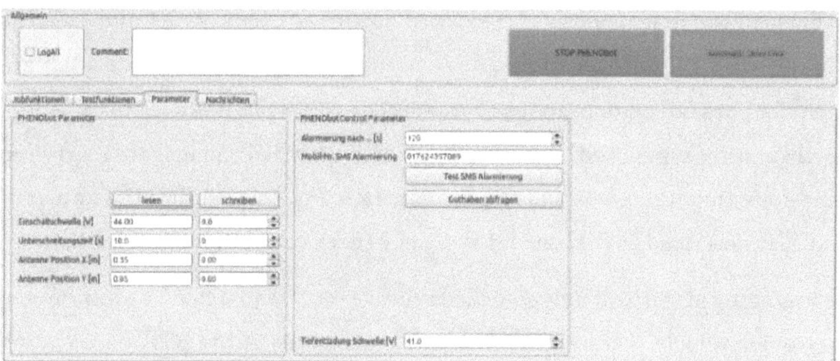

Fig. 79 Parameters tab for setting vehicle and program parameters.

Four vehicle parameters can be configured in the *Parameters* tab (Fig. 79). The *Switch-on threshold* value defines the voltage threshold in volts, which must be undershot for the *undershoot* time in seconds in order to switch on the generator.

The *Antenna position* fields specify the offset of the GPS position antenna in the X and Y directions relative to the center of the vehicle. These vehicle parameters can first be read from the vehicle microcontroller and then configured.

A lower voltage limit can be defined in the *Deep discharge threshold* field to prevent deep discharge of the batteries in the event of a generator malfunction (e.g. lack of fuel). If the voltage falls below this threshold, job execution pauses automatically (Fig. 76) and an SMS alarm is sent. The timeout threshold after which job execution is paused and an SMS alarm is triggered can be defined in the *Alarm after...* field. The telephone number for receiving the SMS alert can also be specified in this area. If a prepaid card is used to send the SMS, the remaining credit on the card can be checked using the *Check credit* button. A commercially available UMTS Internet stick was connected to the navigation computer to send the SMS and made uniquely identifiable using the udev rule. The SMS is sent via the command line application *Gammu* from the PHENObot Control application. The credit query was implemented by sending a USSD code. The status messages of the SMS alarm and the telegram for triggering the FMD are displayed as a text log in the *Messages* tab.

4.6 Navigation

The integrated route planner of the carrier vehicle is controlled by the corresponding libraries provided by the manufacturer KMS. The coordinates must be available in two dimensions as UTM coordinates (see Chapters 2.1 and 4.4). The coordinates with the associated `drvTo command` and a speed specification are sufficient for pure navigation. In the course of the test drives, it proved to be helpful for further processing and evaluation of the recorded images to carry an identification number and assign the image name. For this reason, the format of the text files described in chapter 4.4 was defined for controlling the PHENObot.

At the beginning of the test, only so-called relative coordinates could be interpreted by the carrier vehicle. These referred to the coordinates of the RTK base station and were displayed as grid coordinates in the unit meter in a right-pointing and up-pointing coordinate system. In order to be able to process the JKI Geilweilerhof stock information available as UTM coordinates, this functionality (UTM) was added to the vehicle library. However, it was only possible to select WGS84 as the datum format, which is why a calibration analogous to the method in Chapter 3.6.1 was necessary. The position of the PHENObot base station was measured in UTM/ETRS89 using an external RTK system. The coordinates determined were transferred to the base firmware using a configuration tool. Fig. 80 shows the status of the base station after connecting it to a computer via USB and calling up the tool. The number of satellites received, the voltage of the power supply, the longitude/latitude position in decimal degrees and the current elevation mask in degrees are displayed. In addition, the two status LEDs provide information about the general GPS status and the transmission of a correction signal. Fig. 81 shows the settings dialog in which the reference position can be entered after determination with an external system. When entering the altitude, the undulation must be taken into account; the WGS altitude is entered. The new parameters are communicated to the base using the *Send* and *Save* buttons. As long as the entered position and the position determined via the GPS module differ by less than 2 m, the firmware uses the manual reference position.

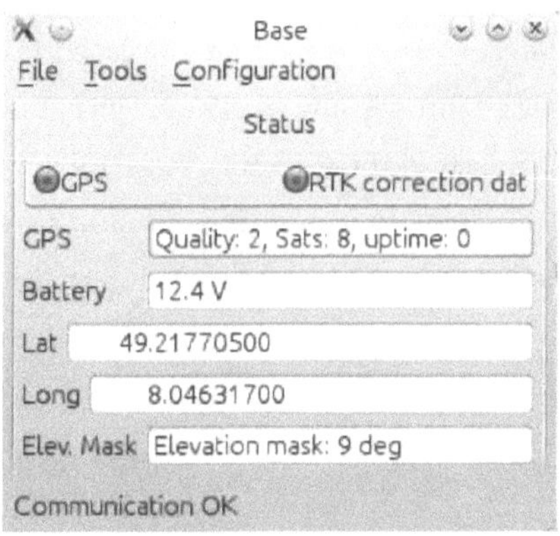

Fig. 80 Firmware RTK base status display PHENObot.

Fig. 81 Manual setting of the reference position the RTK base PHENObot.

During the first test drives in the courtyard of the Institute of Technology, the target points were set manually in the GIS software GRASS GIS. This made it possible to make changes to the attribute table of the linked database graphically and using command line commands. Columns corresponding to the job file structure were inserted into this attribute table and provided with data (point sequence and commands) (Fig. 82).

Fig. 82 Definition of a test course during the first test drives with data from the attack table: commands (`cmds`*) and point sequence (*`success`*).*

The points were placed using the GRASS GIS digitizer tool. In order to be able to place these accurately, a section of a georeferenced aerial image from the geoportal of the state of Hesse (Hessenviewer) was inserted into the project. The origin of the coordinates was set to the position of the RTK base station in order to relate it to the relative coordinates of the carrier vehicle. The vector map generated in this way was exported to a text file using the command line command `v.out.ascii`:

```
v.out.ascii input=vektorkarte@mapset columns=reihenfolg,cmds |
sort -t "|" -k 4 -g > /home/phenobot/jobfiles/beipspiel.txt
```

The coordinates and the columns `reihenfolg` and `cmds` are also output and sorted in the file according to the column `reihenfolg` (column 4). In this form, the generated text file can be read and used by the control program as a job file.

A workflow was developed to convert the available stock coordinates into usable job files.

4.6.1 Import

The `v.in.ascii` command can be used to import text-based file formats into GRASS GIS as vector maps. The stock coordinates must therefore be provided either as a text file or as a csv file. A schema for creating these source files was also created according to the job file structure. The csv type was selected as the format.

Structure of source file with stock coordinates:

1. Column right value (x, Easting),
2. Column superscript (y, Northing),
3. Column StockID (1st point as index S, last point as index E),
4. Sequence column, running from 0 to n,
5. Column ID for recognition 1st (0) and last point (1), the remaining levels remain empty,
6. Bend direction detection column: with direction of travel = 1, against direction of travel = 2, center (no bend) = 0.

Further specifications:

- Coordinates with 3 decimal places,
- Replace comma with decimal point as decimal separator,
- Use | as the separator between the columns,
- In the Easting column, remove the 32 as a zone prefix,
- Do not use column headings,
- One file for each row,
- Naming:

`P[Parcel number]R[Row number].csv`→ Example: `P23R1.csv`.

Extract from a coordinate source file:

```
430471.952|5452213.893|23-001-S|0|0|0
430471.952|5452213.893|23-001-001|1||2
430472.094|5452212.648|23-001-002|2||2
430472.149|5452211.469|23-001-003|3||2
430472.215|5452210.198|23-001-004|4||2
430472.286|5452209.042|23-001-005|5||2
430472.335|5452207.825|23-001-006|6||1
430472.448|5452206.013|23-001-007|7||2
430472.556|5452204.846|23-001-008|8||2
430472.609|5452203.713|23-001-009|9||2
430472.638|5452202.425|23-001-010|10||2
...
430476.935|5452122.318|23-001-E|73|1|0
```

Example of the import command:

```
v.in.ascii input=/home/phenobot/Koordinaten/Geilweilerhof/P23/P23R1.csv
output=P23R1 x=1 cat=0
```

This command reads the source file into GRASS GIS as a vector map and creates a linked database with an attribute table.

1.1.2 Processing

The vine coordinates read in must then be processed further. The carrier vehicle cannot use the coordinates directly, as the lane must be in the middle of the alley between the rows of vines. First of all, it is necessary to know the orientation of the row. To do this, a line was drawn manually on the GRASS GIS GUI through the first and last point of the row in question and its azimuth was derived. This allowed trigonometric calculation of transformation parameters for shifting the coordinates to the center of the lane with knowledge of the row spacing. In the same way, the displacements of the turning points in the headland and the stopping points in the middle of the bend were determined as a function of the bending direction.

Commands for the transformation (`$NAME` replaces the series name P23R1):

```
v.transform input=$NAME@RLP output=${NAME}drv xshift=0.9986
yshift=0.0523;
v.edit map=${NAME}drv@RLP tool=move move=-0.1832,3.4952 where=check=0;

v.edit map=${NAME}drv@RLP tool=move move=0.1832,-3.4952 where=check=1;
v.edit map=${NAME}drv@RLP tool=move move=-0.0065,0.1248 where=dir=1;
v.edit map=${NAME}drv@RLP tool=move move=0.0065,-0.1248 where=dir=2;
```

This creates the vector map P23R1drv in the middle of the lane, taking into account the bending direction, and places the turning points at a distance of 3.5 m from the last floor in the headland. The column names also need to be named. The `v.db.renamecol` command is used for this purpose. The `v.db.addcol` command is used to add a column for the commands, which is described with the commands to be executed using `the v.db.update command.`

4.6.3 Export

The points of the generated vector map for the relevant row are exported using the `v.out.ascii` command:

```
v.out.ascii input=${NAME}drv@RLP columns=id,reihenfolg,cmds | sort -t
"|" -k 5 -g > /home/phenobot/geilweilerhof/P23/${NAME}.txt;
```

4.6.4 Workflow

In order to minimize the manual intervention required when processing the stock coordinates into job files and the associated potential errors, a bash script for automated output was developed during the test phase. The basic process is shown here. During semi-automatic processing, GRASS GIS was always started with the GUI and then the individual commands were entered in the command line instance running in GRASS. For full automation, a surrounding script was created that starts GRASS as a pure command line instance and the actual transformation script as an environment variable:

```
#!/bin/bash
export GRASS_BATCH_JOB=${HOME}/transform.sh
chmod u+x ${HOME}/transform.sh
grass -text ${HOME}/grassdata/UTM_Hessen_RLP/RLP
unset GRASS_BATCH_JOB
```

This starts the transformation script (Fig. 83) and first checks whether a GRASS GIS instance is actually running. Then queries are made in user dialogs and paths to source files and target folders are defined. During the intermediate steps, a number of files are created and stored in a temporary folder, the contents of which are deleted when the script is finished. The way the script works is deliberately destructive; all existing target files with the same name are overwritten.

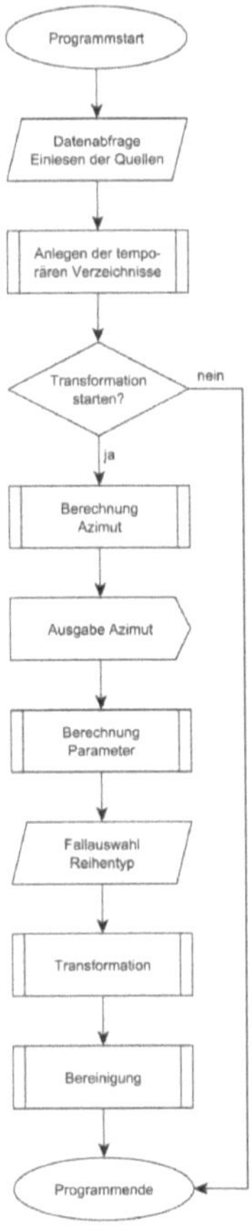

Fig. 83 Transformation script sequence.

There are different transformation scenarios depending on the processing direction and the recording side of the vine line. The PHENObot can only take pictures on the right-hand side when viewed in the direction of travel. In viticultural practice, it is quite common to alternately plant the alleys between the rows of vines. Due to the increased weather resistance of the greening, it is often only the greened alleys that are used for driving. For this reason, the recording sides and thus also the shift of the cane coordinates to the middle of the alley differ with every second row. A distinction was made between even and odd rows based on the naming scheme at JKI Geilweilerhof. Furthermore, the direction of the traffic and the respective position of the lane to be used are decisive. This results in eight possible cases for shifting the coordinates, taking into account the row type (Fig. 84).

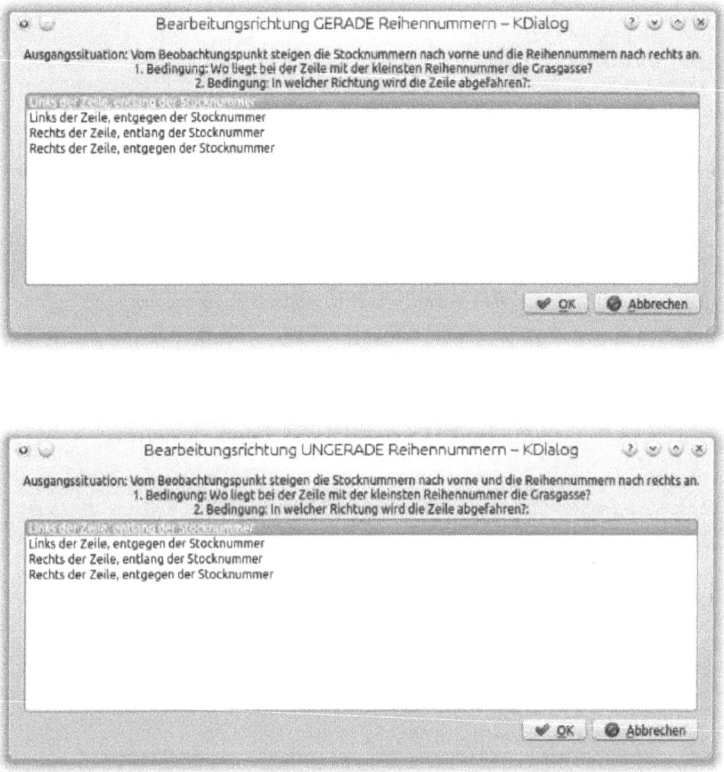

Fig. 84 Case query for transforming the floor coordinates. Top: Even row numbers. Below: Odd row numbers.

Four cases are listed for each of the two main distinctions, even and odd. When answering the question, the user assumes that he is standing in front of the plot in question and that he will find ascending floor numbers in the line of sight and ascending row numbers to his right. Then, for each type, the user is asked on which side of the row with the smallest index the green alley is located and whether it is necessary to drive along or against the ascending floor number.

After reading in the source files, you are asked which commands are to be executed at each breakpoint (Fig. 85).

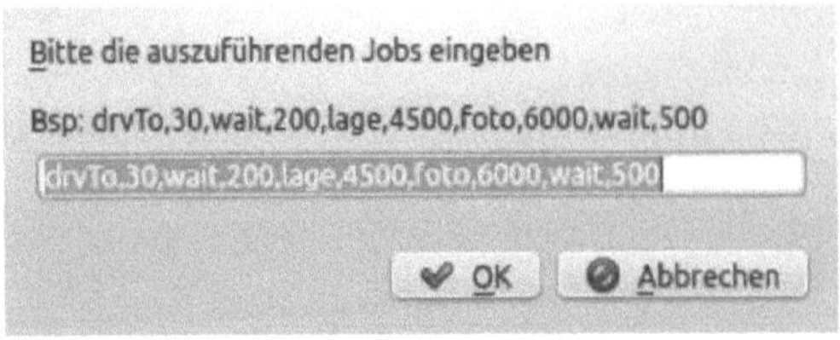

Fig. 85 Specification of the commands to be executed (jobs).

During the test phase, the time-optimized sequence of the image acquisition process with levelling was determined empirically. This sequence with waiting times was inserted into the script as a preselection so that the user only has to confirm.

In the next step, the azimuth is calculated by the script as the average value of all vine rows read in:

```
for FILENAME in `cat $DATFILE`; do
NAME=${FILENAME%.csv};
head --lines=1 ${DIR}/${FILENAME} > ${TMPDIR}/${NAME}linestart.txt;
tail --lines=1 ${DIR}/${FILENAME} > ${TMPDIR}/${NAME}linelast.txt;
paste -s ${TMPDIR}/${NAME}linestart.txt ${TMPDIR}/${NAME}linelast.txt
> ${TMPDIR}/${NAME}lines.txt;
v.in.lines input=${TMPDIR}/${NAME}lines.txt output=${NAME}lines;
v.category input=${NAME}lines@${MAPSET} output=${NAME}line@${MAPSET}
option=add;
v.db.addtable map=${NAME}line@${MAPSET};
v.db.addcol map=${NAME}line@${MAPSET} columns="azimut double
precision";
v.to.db map=${NAME}line@${MAPSET} type=line option=azimut
units=degrees columns=azimut;
db.out.ogr input=${NAME}line dsn=${DIRDIR}${NAME}line.csv format=CSV
g.remove vect=${NAME}line@${MAPSET},${NAME}lines@${MAPSET};
done
```

Lines are drawn through the start and end points and their angles are output in a csv file. These individual files are read out, the value for the orientation is appended step by step to another text file, the mean value is calculated from the individual values and the difference to the north direction is output (Fig. 86):

```
ls ${DIRDIR} > "$TMPDIR"dirdat.lst
DIRDATFILE="$TMPDIR"dirdat.lst
#Mit dieser Schleife wird der Azimut jeder Zeile aus der exportierten
csv ausgelesen und an eine Textdatei angehängt.
for FILENAME in `cat $DIRDATFILE`; do
DIRNAME=${FILENAME%.csv};
cat ${DIRDIR}${DIRNAME}.csv | head -21 | tail -11 | cut -d\, -f2 >>
${DIRDIR}/dir.txt
done

#Es wird der Mittelwert der aufgelisteten Azimut-Werte ausgerechnet
und die Differenz zur Nordrichtung (180°) berechnet
MEAN=`awk '{a = a + $1} END {print a / FNR}' ${DIRDIR}/dir.txt`
NORTH=180
GRAD=`echo "scale=2 ; $NORTH-$MEAN" | bc`
```

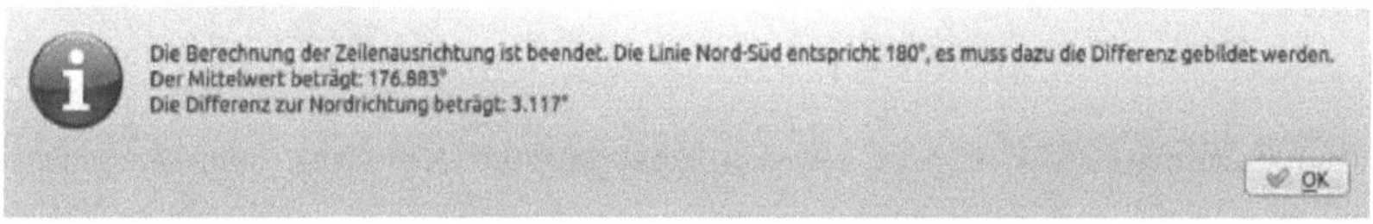

Fig. 86 Output of the azimuth and the difference to the north direction.

The parameters line width, correction of the break point in the middle of the sheet and the distance of the turning point are then queried in a dialog box (Fig. 87).

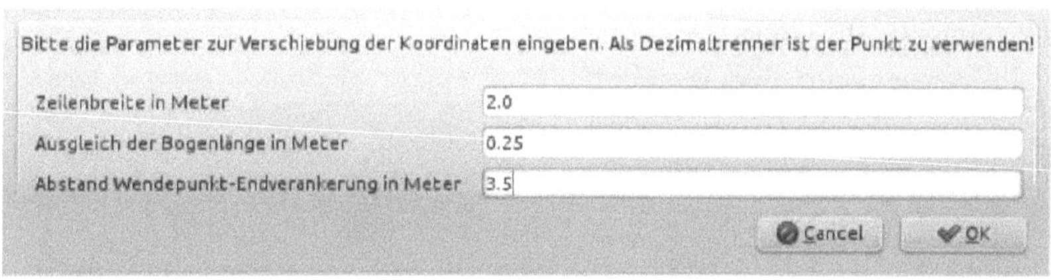

Fig. 87 Parameter query for calculating the transformation parameters.

These values are stored in variables and are included in the calculation of the transformation parameters:

```
#Trigonometrische Berechnung der Transformationsparameter
varRAD=`echo "scale=10 ; a(1)/45" | bc -l`
HZEILE=`echo "scale=3 ; $ZEILE/2" | bc`
XSHIFT=`echo "scale=4 ; c($GRAD*$varRAD)*$HZEILE" | bc -l`
YSHIFT=`echo "scale=4 ; s($GRAD*$varRAD)*$HZEILE" | bc -l`
CXSHIFT=`echo "scale=4 ; c($GRAD*$varRAD)*$CHECK" | bc -l`
CYSHIFT=`echo "scale=4 ; s($GRAD*$varRAD)*$CHECK" | bc -l`
DXSHIFT=`echo "scale=4 ; c($GRAD*$varRAD)*$BOGEN" | bc -l`
DYSHIFT=`echo "scale=4 ; s($GRAD*$varRAD)*$BOGEN" | bc -l`
```

Due to inaccuracies when measuring the vines with the RTK measurement system (e.g. GPS antenna not perpendicular to the measuring point), more pronounced yaw movements around the vertical axis were observed during the test phase when approaching several points sequentially. This deviation from the linear track through the Rebgasse was accompanied by misalignments of the FMD to the foliage wall. For this reason, the line generated for the azimuth calculation is used in the script as the straightening axis of the point chain. The perpendicular distance to this line is determined from each point and saved as an attribute in the database. These values are used to generate a new vector map whose points now lie on the generated line and thus result in a smoothing of the lane:

```
v.distance --o to=${NAME}lines from=${NAME}@${MAPSET} out=connectors
upload=to_x,to_y col=x,y;
v.db.select -c ${NAME}@${MAPSET} col=x,y,id,reihenfolg,check,dir,a,b
file=${TMPDIR}/${NAME}select.txt;
v.in.ascii input=${TMPDIR}/${NAME}select.txt output=${NAME}l x=1
cat=0;
```

The transformation and export process was adopted from the semi-automatic version during script development, only the parameters were replaced by the variables.

A transformation loop was written as a separate function for each of the eight cases. Depending on the query result, when the script is executed for the even and odd row numbers, the respective

function is called:

```
#Fallauswahl und Funktionsaufruf
case "$ORIENTG" in
  0) left_along_g;;
  1) left_con_g;;
  2) right_along_g;;
  3) right_con_g;;
esac

case "$ORIENTU" in
  0) left_along_u;;
  1) left_con_u;;
  2) right_along_u;;
  3) right_con_u;;
esac
```

Before the script ends, the temporarily created files and folders are cleaned up.

As a result, the user receives a folder with job files for each reb line read in. The identical individual commands are executed at every point, including at the turning points. If this is not desired, the individual files must be changed manually. The generated job files can be loaded into the task planner via the *JobFile* button of the PHENObotControl 1.0 application.

As a rule, however, several rows of vines have to be traversed sequentially during scoring. The *ProjectFile* function is available in PHENObotControl for this purpose. This reads out text files in which the paths to individual job files are listed. In this way, projects can be constructed modularly from different rows, device functions and approach paths. An example of a device function is the job file for switching on the generator with the command `generator,1`.

Extract from the project file P23.txt

```
/home/phenobot/Projekte/geilweilerhof/P23/P23R1.txt
/home/phenobot/Projekte/geilweilerhof/P23/P23R2.txt
/home/phenobot/Projekte/geilweilerhof/P23/P23R3.txt
/home/phenobot/Projekte/geilweilerhof/P23/P23R4.txt
/home/phenobot/Projekte/geilweilerhof/GeneratorOn.txt
/home/phenobot/Projekte/geilweilerhof/P23/P23R5.txt
/home/phenobot/Projekte/geilweilerhof/P23/P23R6.txt
/home/phenobot/Projekte/geilweilerhof/P23/P23R7.txt
/home/phenobot/Projekte/geilweilerhof/P23/P23R8.txt
/home/phenobot/Projekte/geilweilerhof/GeneratorOn.txt
/home/phenobot/Projekte/geilweilerhof/P23/P23R9.txt
/home/phenobot/Projekte/geilweilerhof/P23/P23R10.txt
```

A bash script was also written to create these project files:

```
#!/bin/bash
if ! DIR=$(kdialog --title "Bitte das JobList-Verzeichnis wählen!" --
getexistingdirectory ${HOME}/Projekte/); then
exit;
fi

if ! PROJECT=$(kdialog --title "Projektname eingeben" --inputbox
"Bitte den Projektnamen eingeben!"); then
exit;
fi

if ! SORTDIR=$(kdialog --title "Sortierrichtung" --radiolist "Auswahl
der Sortierrichtung:" 0 "Vorwärts (Ordnerreihenfolge)" on 1
"Rückwärts" off); then
exit;
fi

function out_along {
find / -path "${DIR}/*.txt" 2>/dev/null | sort -t "/" -k 5 --version-
sort > ${HOME}/Projekte/${PROJECT}_along.txt;
}

function out_con {
find / -path "${DIR}/*.txt" 2>/dev/null | sort -t "/" -k 5 --version-
sort -r > ${HOME}/Projekte/${PROJECT}_reverse.txt;
}

case "$SORTDIR" in
  0) out_along;;
  1) out_con;;

esac
```

The contents of the job folder are read out here and output either alphabetically or in the opposite direction as a project file.

1.1.5 Editing in QGIS

The GIS program QuantumGIS was used to visualize and check the results of the coordinate transformation. It is also used to evaluate the test and measurement runs.

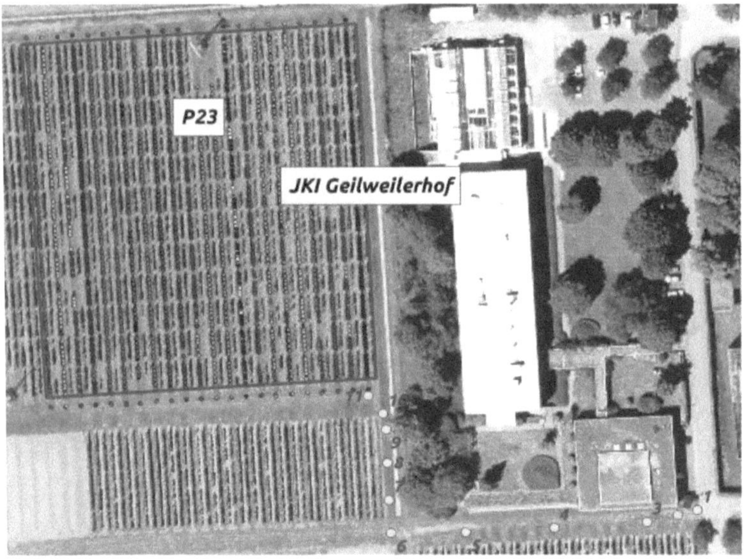

Fig. 88 Paths P23 with access route (points 1-11).

Another use case was the generation of the access route job files. Fig. 88 shows the generated paths in plot 23. The yellow points were manually placed in a vector layer of the GRASS GIS project using the QGIS digitizer tool. Using another script, it is possible to export these manually generated vector maps to job files. The source map selection can be made in user dialogs. The selected map is then exported using the `v.out.ascii` command. It is possible to select the point sequence.

The special feature of QGIS is that it can be linked to GRASS GIS. This is done with the GRASS extension package. On the one hand, all layers of a GRASS GIS project can be directly integrated into a QGIS project. On the other hand, several GRASS GIS tools are available in QGIS, with which the integrated GRASS layers can be edited externally in QGIS.

4.7 Virtual machine

The processing and transformation of the stock coordinates is largely carried out with bash scripts of a Linux system. The GRASS GIS and QGIS installations used are also based on this system. In order to remain independent of the user's operating system and computer equipment, and to avoid further installation and configuration work, the system used was set up in a virtual machine (VMware Player 6.0) (Fig. 89). This means that it can be used in most guest systems regardless of the operating system.

Fig. 89 Virtual machine for transforming and configuring the PHENObot RTK base station

The virtual machine has been fully configured and can be used directly. It is possible to keep a backup copy of the original state and restore it if required. Kubuntu 12.04.4 LTS was used as the operating system. All necessary programs and the scripts mentioned were installed and extensively tested.

4.8 Test drives

All test drives described were carried out with the already modified test carrier based on the Neosat Alpin. All findings on the Neo Alpin can be found in chapter 4.1.

4.8.1 Test drives Geisenheim

When satellite-based navigation became available in spring 2013, the first tests were carried out with the carrier vehicle

Fig. 90 First test runs with Neosat Alpin in the courtyard of the Institute of Technology.

Fig. 90 shows the Neosat Alpin in the courtyard of the Institute of Technology at Geisenheim University. A test course (Fig. 82) was defined and run using relative coordinates. During the first run, the turning points were marked with pylons at the pivot point on the side of the inner crawler track. The absolute deviation measured against the pylons was less than 50 mm after five passes.

Fig. 91 First test runs with Neosat Alpin in the courtyard of the Institute of Technology. Circle: Position of the RTK base.

The RTK base station was mounted on the roof of the institute building for testing (Fig. 91). During tests, shadowing effects and dependencies on the diurnal cycle of satellite availability could be detected in the courtyard. The number of received satellites (SV) on the rover on open terrain was between eight and ten SV per module at times with the greatest availability. At driving positions, as shown in Fig. 91, between six and seven SVs were received at normal reception times. In times of low availability, the number of SVs was sometimes below fanf. The journey was then interrupted. A lower threshold of at least fanf available SVs for automatic driving operation was defined in the vehicle library. The planning application PlanningOnline from Trimble was used to plan the test and measurement runs (Fig. 92). At the time of recording (March 7, 2013, 13:35) of Fig. 91, nine SVs were visible at an elevation of 10° at the Geisenheim site according to PlanningOnline (Fig. 92). There are no automatically logged numbers of available SVs for this trip, as this function was not yet implemented in the control application at this time. According to manual recordings, the number in the range shown was between six and seven SV.

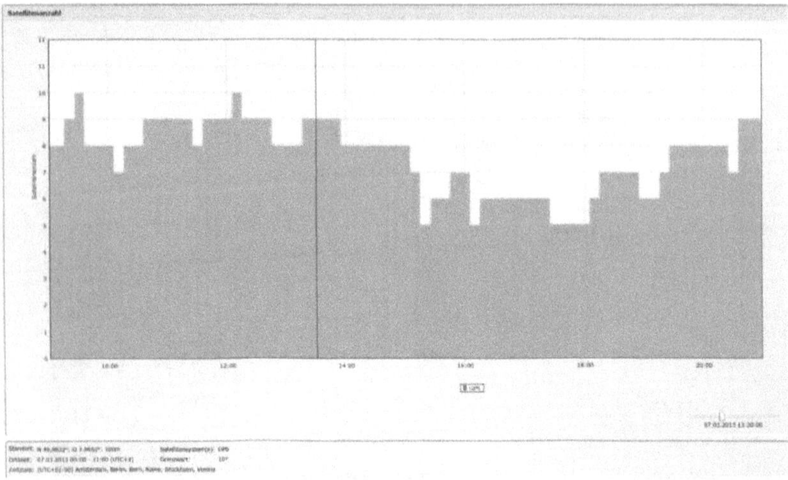

Fig. 92 Predicted visibility of the GPS SV in Geisenheim on March 7, 2013, 9:00-21:00 with PlanningOnline (http://www.trimble.com/GNSSPlanningOnline/#/NumSats).

In times of very good availability, the correction signal was lost several times during the test runs. Documented examples of this are August 31 (16:46) and September 1, 2013 (16:41). At these times, the carrier vehicle came to a sudden stop while driving. The log file of the control application indicates nine available SVs for the rover. No such automatic output with documentation was available for the GNSS module of the base station. When looking at the visible SVs in PlanningOnline, an increase to ten SVs can be seen for this time period (Fig. 93). However, the number of available SVs could be read on the display of the base station and documented manually. The number was mostly ten, briefly also eleven SVs.

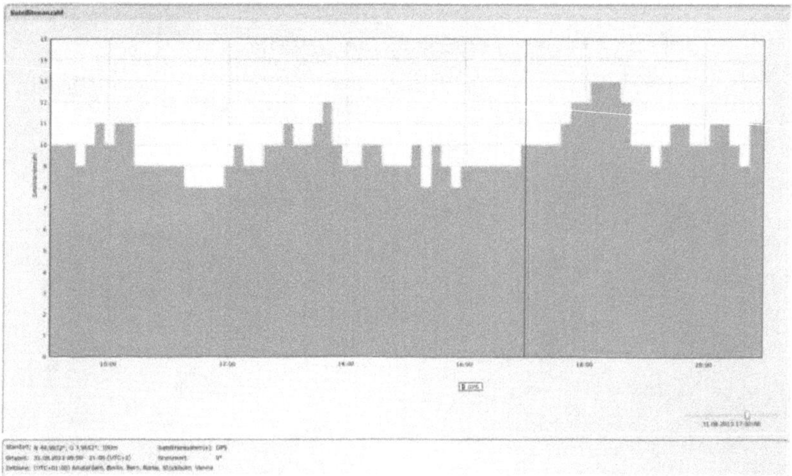

Fig. 93 Predicted visibility of the GPS SV in Geisenheim on August 31, 2013, 9:00-21:00. Elevation set to 0° as the base had no visible shading.

PlanningOnline (http://www.trimble.com/GNSSPlanningOnline/#/NumSats).

The test drives in 2013 and 2014 were mainly used to check, debug and optimize the control application, navigation and workflow. For the PHENObotControl control application, 28 major version changes are documented up to the final release 1.0 (see PHENObotControl version history in the electronic appendix). The revision of the vehicle library for the use of UTM coordinates was completed in August 2013. The offset between the ETRS89 and WGS84 datum formats was compensated for manually using two offset parameters in the control application until the external RTK surveying system from Trimble was recommissioned in spring 2014. Once the location of the RTK base station was calibrated to the ETRS89 datum format, this manual offset became obsolete.

Fig. 94 Tracks before (left) and after (right) calibration with the ETRS89 date format (identical job file).

Fig. 94 shows the displacement of the track after calibration. In the following test runs, this could be checked and reproduced in the test facility of the Institute for Vine Breeding. The calibration of the vine coordinates and the calibration of the RTK base were carried out with the same measurement system. As a result, the test tractor was able to enter the aisle exactly in the middle after the transformation of the cane coordinates using the correct aisle width (Fig. 95).

Fig. 95 PHENObot in the test facility of the Institute for Grapevine Breeding, Hochschule Geisenheim University. Correct track after calibrating the position of the RTK base station.

In the course of these tests, initial tests were also carried out in the Dammerung. The camera illumination beams were used during these trips (Fig. 96). A rough determination of the pure electric range was made. With an average row length of 65 m, an average of six complete rows could be driven with lighting before the generator was automatically switched on at 42 V system voltage. This results in a scoring performance of 300 canes or 702 m^2per battery charge.

Fig. 96 PHENObot during image acquisition with LED illumination bar.

4.8.2 Measuring trips Geisenheim

Fig. 97 shows the number of available satellites as a function of the GPS antenna setting. The measurements were repeated three times in rows 33 (left partial figures) and 34 (right partial figures) of the "vine cultivation" test plot. In all runs, a higher availability was recorded in the antenna setting `ant_up` compared to the low setting `ant_down`. In the upper setting, the number of available SVs was consistent and stable over the measurement period. The vehicle used eight or more SVs for navigation at all times during the measurement. In contrast, the lower setting showed significant fluctuations of up to four SVs. During the second repetition in row 33, the lower threshold of five available SV was reached. All distributions are highly significantly different after statistical testing with the Wilcoxon test for pair differences (Fig. 97).

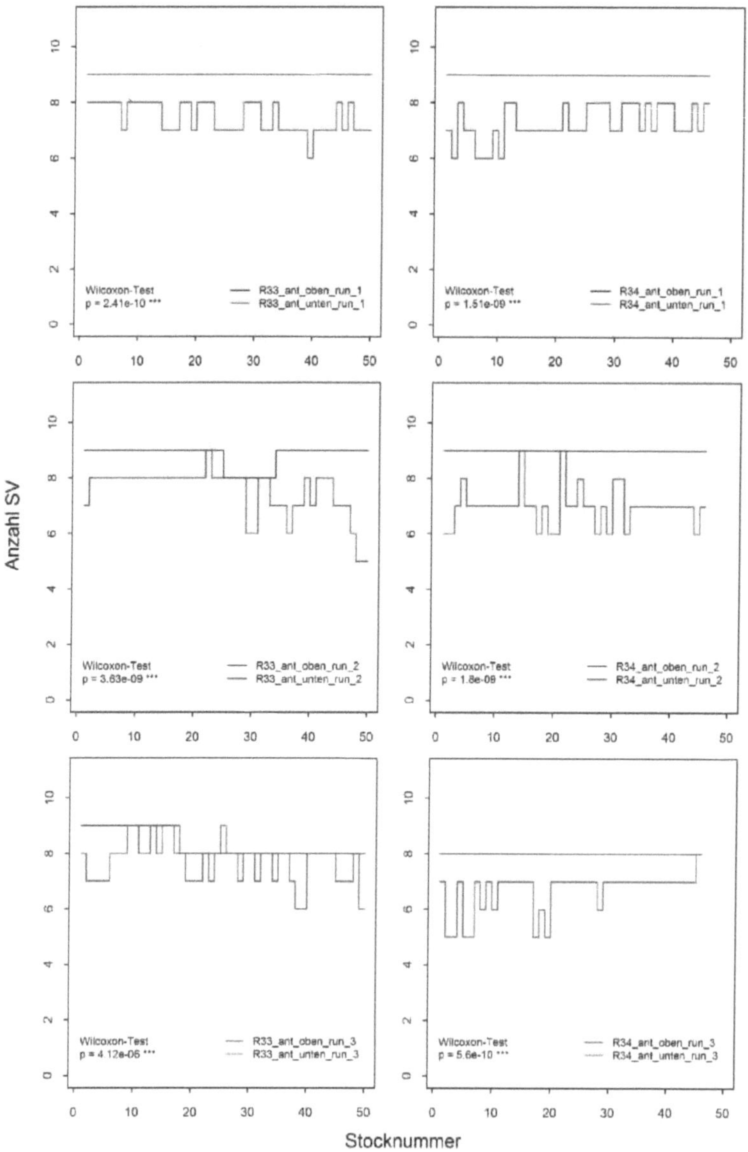

Fig. 97 Measurement of SV availability in the experimental vine breeding facility, HS Geisenheim on September 10, 2014. left: Replicates 1-3 in row 33, Right: Replicates 1-3 in row 34. Trial period 15:07-17:08.

Fig. 98 Choropleth map for visualization of SV availability in row 33, HS Geisenheim breeding trial on 10 September 2014. Left: GPS antenna setting 1280 mm above ground (ant_bottom); Right: GPS antenna setting 2000 mm above ground (ant_top).

A choropleth map was created to visualize the different availability depending on the GPS antenna height setting. Fig. 98 shows the availability limited to the lane between rows 33 and 34, when driving south with the coordinates of row 33 on September 10, 2014. An interpolation of the availability was created from the measurement data of the log file for the lane pools. The left part of the figure shows the measurement during the test drive with the `ant_down` setting. The availability fluctuated between six and eight SV. In the northern part of the test area, the plot shows a slightly higher availability, whereas a lower number of SVs was received in the southern area. The right part of the figure shows the availability during the drive with setting `ant_up`. Here the availability was consistently nine SV.

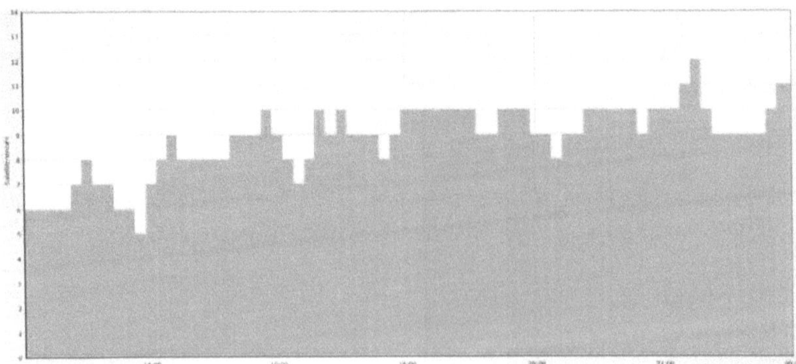

Fig. 99 Forecast of SV availability in Geisenheim on September 10, 2014, 12:00-0:00. Trial period 15:07-17:08 with PlanningOnline (http://www.trimble.com/GNSS-PlanningOnline/#/NumSats).

To assess the potential availability of SV, a prediction of the constellation was carried out using the planning tool PlanningOnline (Fig. 99). For the test period from approx. 15:00-17:00, there was a trend towards an increase in the number of receivable SVs. The predicted fluctuation range during this period was six to ten SV. The tests with the `ant_oben` setting were carried out at the beginning of the period.

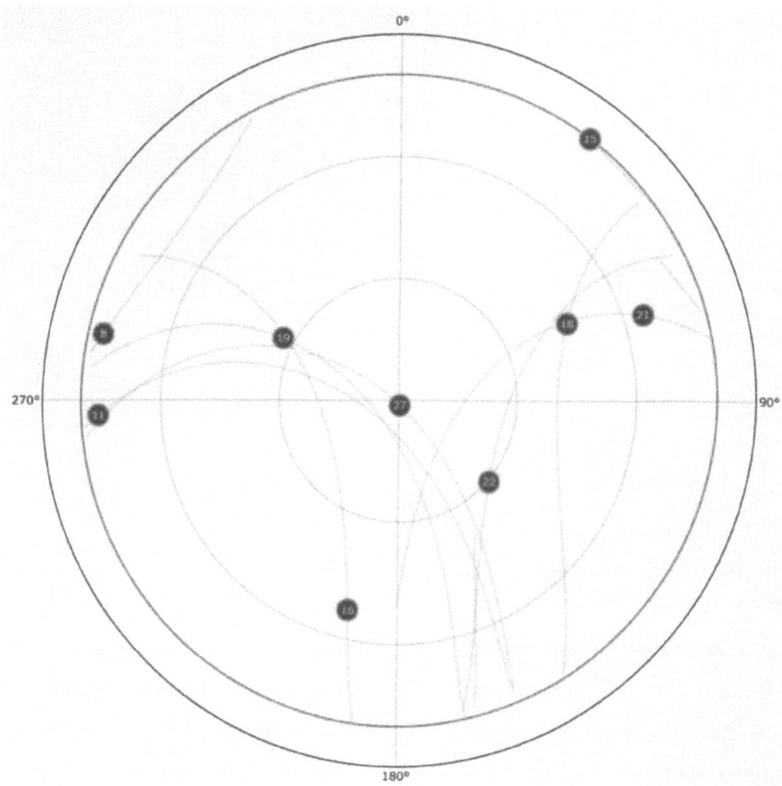

Fig. 100 Illustration of the SV constellation on September 10 in Geisenheim at 4 p.m. with PlanningOnline (http://www.trimble.com/GNSSPlanningOnline/#/NumSats).

To characterize the potential shading effects of buildings, vegetation or terrain profile, the SV constellation was investigated at the horizon at the time of the test. At this time, the nine available SVs were mainly distributed on the western, southern and eastern horizon. In the northern sector, only satellite 15 was in the reception range for a short time (Fig. 100).

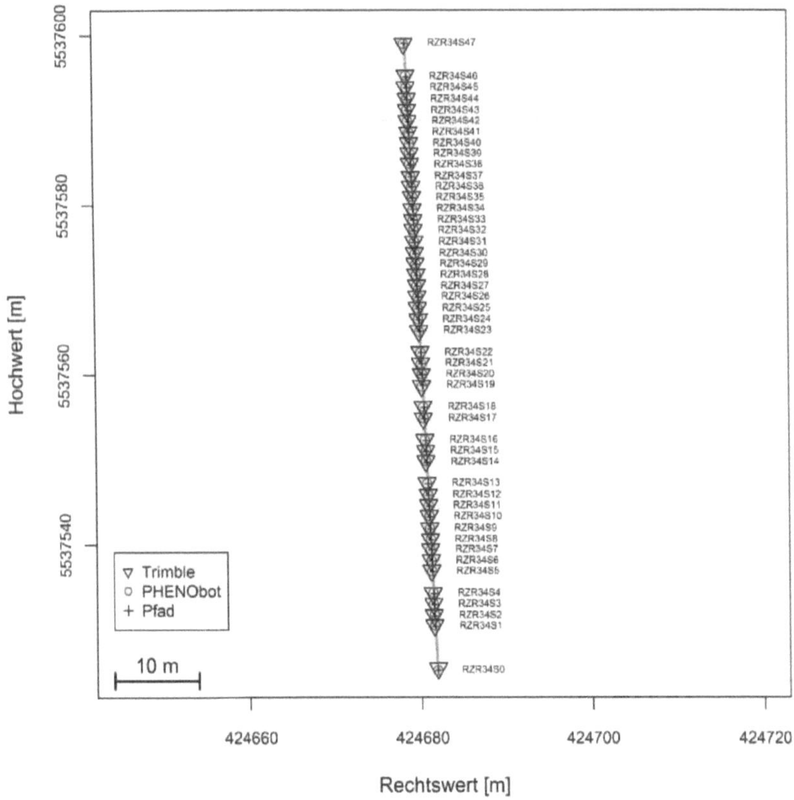

Fig. 101 Overview of the various positions, measurement run on December 2, 2014, row 34, repetition 4, measured values from the Trimble measurement system, PHENObot log file and path file.

The measured values of the accuracy determination using the example of the "Rebenzüchtung" plot, HS Geisenheim, fourth repetition, carried out on December 2, 2014 are shown in Fig. 101 as an overview. The target points of the path are shown as black crosses, the corresponding positions recorded via the internal GNSS system of the PHENObot (red circles) and the Trimble RTK surveying system (black triangles) in the UTM coordinate system. The deviations between the measured values cannot be seen in this overview.

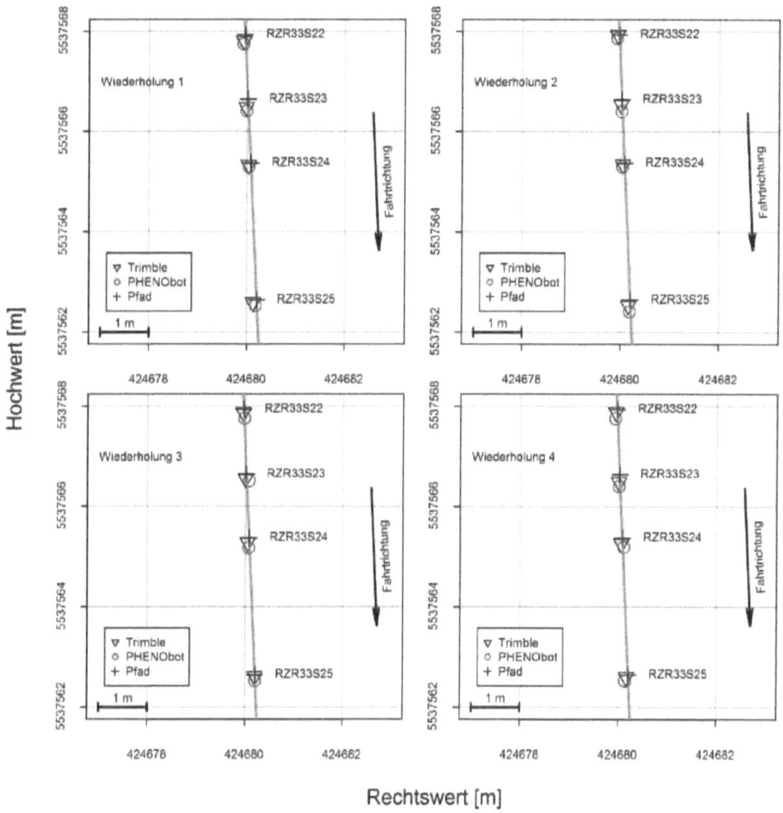

Fig. 102 Detailed view of the various positions, measurement run on December 2, 2014, row 33, repetitions 1-4. Measured values from the Trimble surveying system, PHENObot log file and path file.

The detailed views of the individual repetitions for the points RZR33S22-25 are shown in Fig. 102. For evaluation purposes, the mean errors from the four repetitions per row are examined below.

In the transverse direction perpendicular to the direction of travel, the deviations were small for both measuring systems. Expressed as RMS error, the absolute position accuracy was on average 42 mm (Trimble) and 37 mm (PHENObot). The deviations were significantly greater in the longitudinal direction. Measured with both systems, the vehicle traveled on average 107 and 108 mm beyond the theoretical stopping point. The mean values of both measurement systems did not differ significantly from each other (Table 7).

Tab. 7 Descriptive statistics for comparing the absolute positional accuracy of the PHENObot in row 33. The data basis is the mean RMS errors from 4 runs measured with the Trimble RTK system and the internal system. Positive test for normal distribution with the Kolmogorov-Smirnov test.

	Transversal		***Longitudinal***	
	Trimble	PHENObot	Trimble	PHENObot
Mean value (x) [m]	0,041	0,035	0,095	0,096
Standard deviation (s_x)	0,005	0,002	0,010	0,010
Coefficient of variation (cv)	0,132	0,061	0,108	0,107
Number (n)		4		
Minimum [m]	0,036	0,032	0,086	0,087
Maximum [m]	0,047	0,037	0,105	0,105
P	0,1175		0,9148	
Significance	H0 - not significant		H0 - not significant	

The other descriptive statistical key figures are comparable for both variants. The coefficient of variation shows a slightly larger scatter in the data basis of the Trimble system for the transversal deviation. This larger variation is also shown by the larger box of the box-whisker plot in Fig. 103.

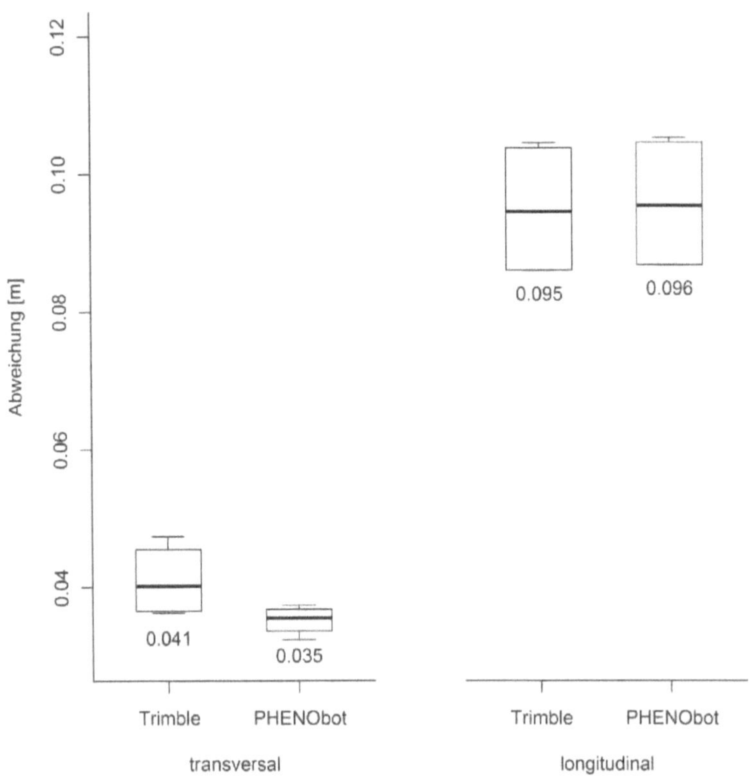

Fig. 103 Representation of the absolute accuracy as mean RMS error in row 33, recorded with 2 RTK systems. Left: Transversal offset to the lane; Right: Longitudinal offset of the stopping point.

When looking at the detailed view of the stopping points RZR34S22-25 when traveling in the opposite direction, position errors are revealed that are approximately two times greater than in row 33. In this row too, the vehicle has traveled beyond the stopping point in the longitudinal direction.

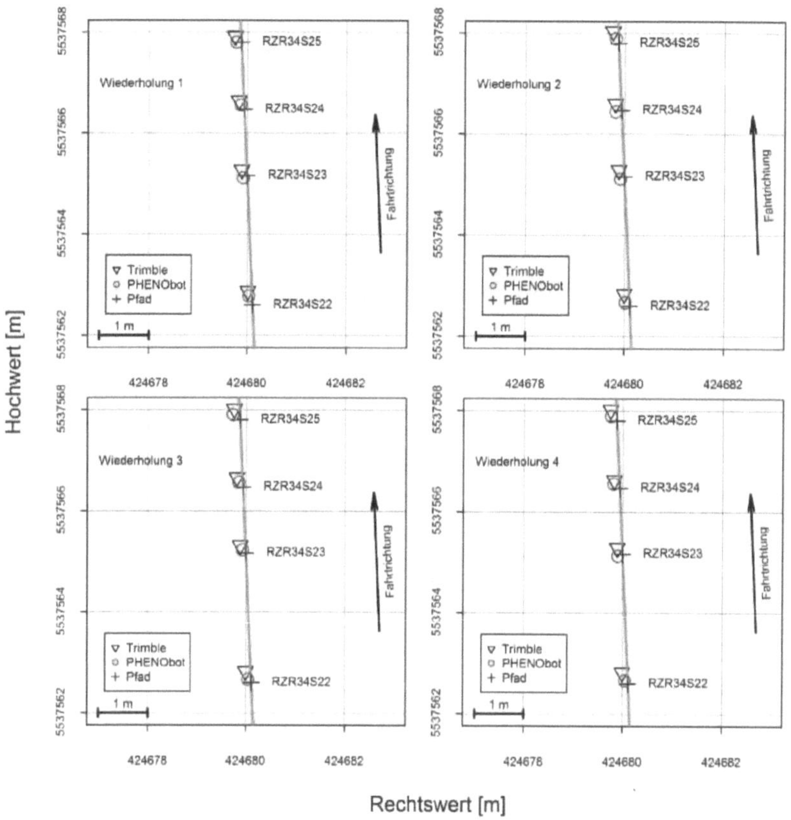

Fig. 104 Detailed view of the various positions, measurement run on December 2, 2014, row 34, repetitions 1-4. Measurement values from the Trimble surveying system, PHENObot log file and path file.

The descriptive statistics (Tab. 8) confirm the observations of the tail view. With the Trimble RTK system, the average transverse deviation from the lane was 97 mm and 188 mm in the longitudinal direction. A deviation of 81 mm transverse and 187 mm longitudinal to the direction of travel was determined from the internally recorded measured values.

Tab. 8 Descriptive statistics for comparing the absolute positional accuracy of the PHENObot in row 34. The data basis is the mean RMS error from 4 runs, measured with the RTK system from Trimble and the internal system. Positive test for normal distribution with the Kolmogorov-Smirnov test.

	Transversal		*Longitudinal*	
	Trimble	PHENObot	Trimble	PHENObot
Mean value (x) [m]	0,097	0,081	0,188	0,187
Standard deviation (sx)	0,004	0,004	0,009	0,009
Coefficient of variation (cv)	0,042	0,054	0,049	0,049
Number (n)		4		
Minimum [m]	0,091	0,075	0,179	0,178
Maximum [m]	0,101	0,085	0,201	0,200
P	0,0019		0,9016	
Significance	HI - significant **		H0 - not significant	

The samples of the RMS errors show identical standard deviations in series 34. The coefficient of variation only shows a slightly larger scatter for the error in the transverse direction in the PHENObot measurement series (Fig. 105). However, the statistical examination of the mean values shows a significant difference. The mean error, calculated according to the internal measured values of the PHENObot, is smaller than that of the RTK system from Trimble.

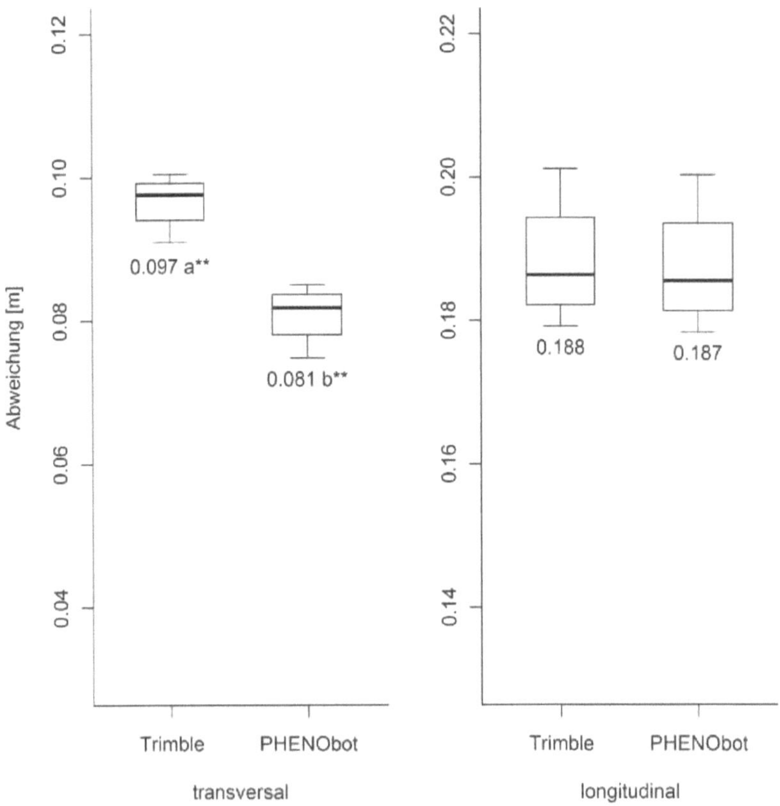

Fig. 105 Representation of the absolute accuracy as mean RMS error in row 34, recorded with 2 RTK systems. Left: Transversal offset to the lane; Right: Longitudinal offset of the stopping point.

To assess the relative position accuracy, the mean RMS errors from four repetitions in a row (R33 and R34) were compared with each other (Fig. 106). To determine the error, the Euclidean distances between the position data of the internal RTK system were determined in the transverse and longitudinal directions. According to combinatorics, a variation without repetition is assumed here with the individual positions as objects (n) and minuend and subtrahend as available places (k): n = 4; k = 2; the following applies:

$$\frac{n!}{(n-k)!} \quad \text{Gl. 4-1}$$

This results in twelve variations for minuend and subtrahend and thus six differences, which were statistically analyzed as a data basis.

Tab. 9 Descriptive statistics for comparing the relative positional accuracy (transverse/longitudinal) of the PHENObot. Data basis are the RMS errors from 4 runs in 2 rows (R33/R34).

	Transversal		***Longitudinal***	
	R33	**R34**	**R33**	**R34**
Mean value (x) [m]	0,038	0,038	0,073	0,074
Standard deviation (sx)	0,005	0,003	0,005	0,007
Coefficient of variation (cv)	0,129	0,08	0,065	0,095
Number (n)			6	
Minimum [m]	0,034	0,033	0,064	0,067
Maximum [m]	0,047	0,041	0,077	0,086
P	0,7853		0,7463	
Significance		H0 - not	t significant	

The mean values of the samples are not significantly different. This means that the null hypothesis is maintained and the transverse and longitudinal deviations in both rows are indistinguishable. At the time of the test, the PHENObot was traveling in the rows with a positional accuracy of 38 mm transversely and 73 or 74 mm longitudinally to the direction of travel. The transverse values in row 33 were subject to slightly greater scatter than in row 34, whereas the opposite was the case in the longitudinal direction (Fig. 106 and Tab. 9).

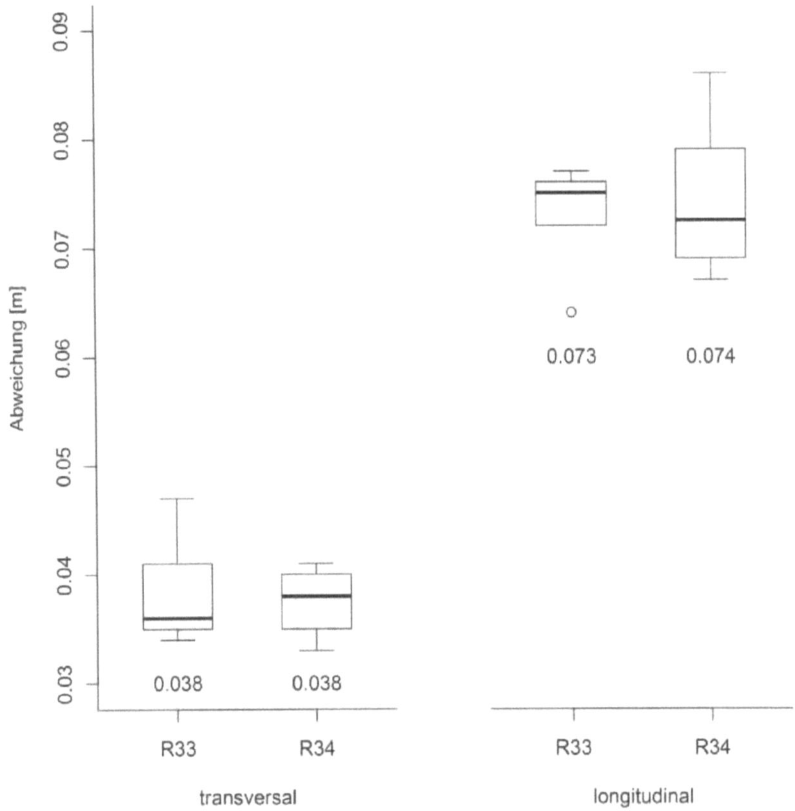

Fig. 106 Relative positional accuracy as RMS error of the mean values in the transverse (left) and longitudinal (right) directions during the test drives on December 2, 2014 (n=4). The mean values of the deviations, based on rows 33 and 34, do not differ significantly (transverse, P=0.7853; longitudinal, P=0.7463).

4.8.3 Test drives Siebeldingen

The tests at JKI Geilweilerhof in Siebeldingen focused primarily on the linking and cooperation between the carrier vehicle and the FMD as well as the processing and correct interpretation of the transferred stick coordinates. The first test runs with a statically mounted FMD were carried out in June 2013 (Fig. 107).

Fig. 107 Trial tower Neosat Alpin with statically mounted FMD in plot 19 during the first trial at the JKI.

The tests were carried out with relative coordinates. For this purpose, reference coordinates were collected on the JKI site with the test vehicle and the external RTK surveying system (white-yellow GNSS antenna) from the JKI company Trimble. With these reference values, the pole coordinates could be transformed into the reference system (relative) of the Neosat Alpin and used. During the first tests, the image acquisition of the MKS was triggered manually. An additional computer was available on the vehicle for this purpose.

In the course of these test drives, a defect occurred in one of the gearboxes for the traction drive (Fig. 108). Both gearboxes were subsequently exchanged by the manufacturer and replaced by more stable designs with a different transmission ratio.

Fig. 108 Gearbox defect on the travel drive due to shearing of the gear flanks.

Following the completion and implementation of the PHENObot, further test drives were undertaken during the vegetation phase in 2014. The software-controlled image drawing, navigation and coordinate processing were examined and optimized. A calibration measurement of the PHENObot's RTK base station with the external RTK surveying system was also carried out at the JKI to adjust the different date formats. In this case, too, no offset adjustments were necessary to correct the travel path. In plots with a larger side slope, an offset of the lane in the order of 100-150 mm to the center of the lane could be determined. Fig. 109 shows the PHE NObot in an alley of plot 23 with a side slope. There was no misalignment of the lane with this size of slope. However, the leveling effect of the camera position control can be clearly seen in the figure. The FMD was thus aligned parallel to the foliage wall. In these tests, the horizontal distance to the foliage wall plane proved to be an uncritical size without too great an effect on the image sharpness.

Fig. 109 PHENObot during testing in plot 23.

Fig. 109 also shows the problem of different lighting due to solar radiation, as already described by the JKI. The left foliage wall is in direct sunlight, whereas the right side is in the shade. For this reason, tests were also carried out in darkness to prepare for the planned measurement runs (Fig. 110). These tests revealed the need for a lighting unit at the front of the PHENObot. For this reason, a bright and efficient LED headlight was fitted to illuminate the paths.

Fig. 110 Test drive in complete darkness in plot 23.

Due to the fully developed foliage wall in this vegetation phase (BBCH 75-85), shadowing could be detected during satellite reception when the two GPS antennas were in the lower position at 1 280 mm above ground. This was not detectable in the upper position (2,000 mm).

4.8.4 Measuring trips Siebeldingen

The evaluation of the number of available SVs as a function of the GPS antenna setting was carried out in Siebeldingen in the same way as in Geisenheim. For this purpose, the first two rows in plot 23 were examined on August 29, 2014. Fig. 111 shows the results. In both rows, the availability was significantly greater in the `upward` setting than in the `downward` setting. The availability level was between four and eight SV in the lower range, with a slightly higher level between floors 1-25 and a lower level between floors 26-71.

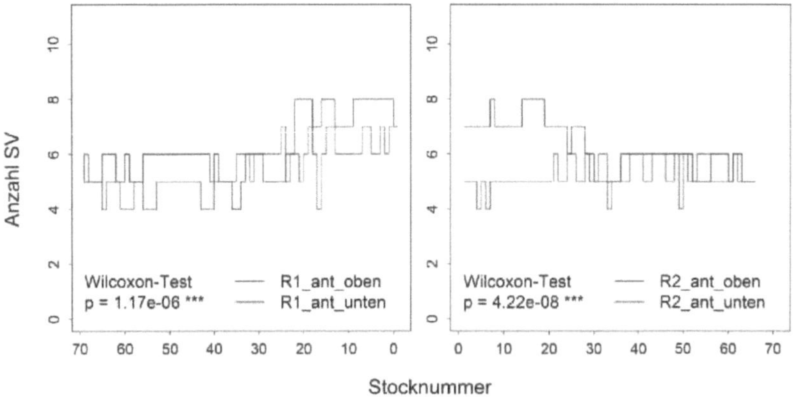

Fig. 111 Measurement of SV availability in the test facility P23, JKI Geilweilerhof on August 29, 2014. left: row 1, right: row 2: Row 1, Right: Row 2, experimental period 11:30-13:00.

Fig. 112 shows this pattern in the form of a choropleth map. In the northern part of the plant with cane numbers 1-25, the availability is medium to good with 5-7 SV (orange-light green) in the `ant_down` setting and 6-8 SV (light orange-green) in the `ant_up` setting.

Fig. 112 Choropleth map for visualization of SV availability in row 1, experimental plant P23, JKI Geilweilerhof on 29 August 2014. Experimental period: 11:30-13:00. Left: Setting GPS antennas 1280 mm above ground; Right: Setting GPS antennas 2000 mm above ground.

The building complex with trees in the south-eastern area is remarkable. Looking at the SV constellation in Fig. 113, four SVs (SV2, SV6, SV15, SV17) were flying at low elevation (10°-20°) on the south-eastern horizon at the time of the experiment.

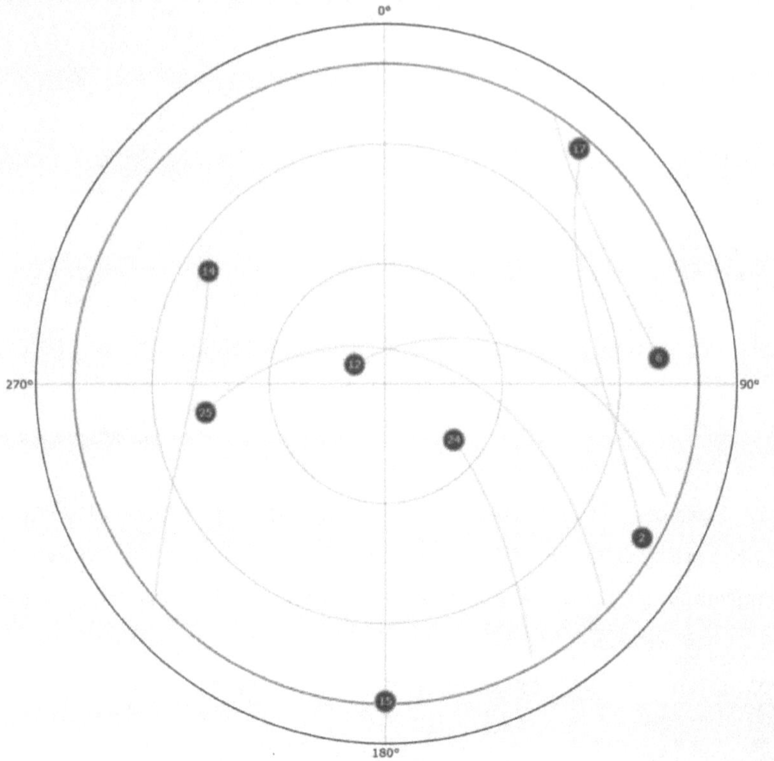

Fig. 113 Representation of the SV constellation in Siebeldingen on August 29, 2014, 09:00-15:00. Trial period 11:30-13:00 with PlanningOnline (http://www.trimble.com/GNSSPlan- ningOnline/#/NumSats).

During the test period (11:30-13:00), an availability of six to seven SV with an elevation mask of 10° was predicted (Fig. 114).

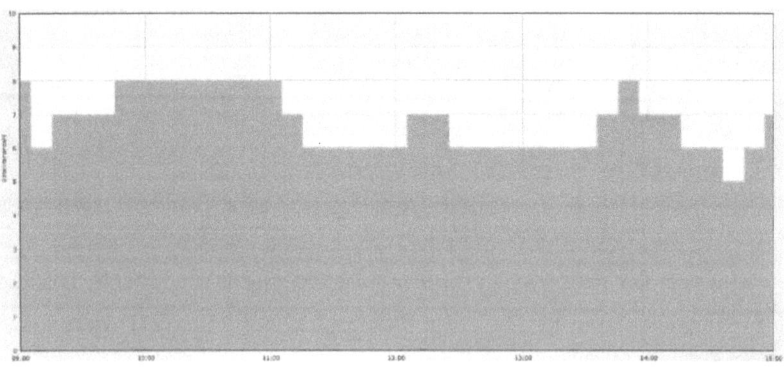

Fig. 114 Forecast of SV availability in Siebeldingen on August 29, 2014, 09:00-15:00. Search period 11:30-13:00

with PlanningOnline (http://www.trimble.com/GNSSPlan- ningOnline/#/NumSats).

The SV availability was recorded in the course of the application trial. Fig. 115 shows the results graphically as a choropleth map in rows 7-43. It should be noted that the map was created using data from two night trips. On the night of September 16-17, rows 1-25 were sampled and on the following night rows 26-43 were sampled.

Fig. 115 Choropleth map for visualization of SV availability in rows 7-43, experimental plant P23, JKI Geilweilerhof on 16 and 17 September 2014. Trial period: 20:00-02:00 each day.

During the test period, there was consistently good to very good availability (yellow-blue, 8-10 SV). Only at some positions (red-orange) were 6 (n=4) or 7 (n=84) SV received. In total, SV availability was recorded and evaluated at 2,279 positions in rows 7-43.

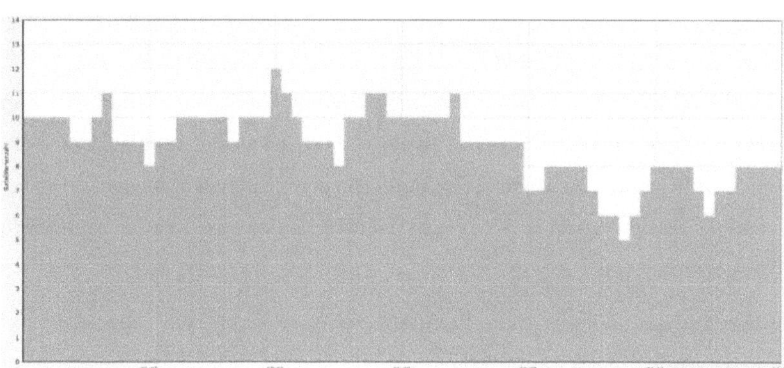

Fig. 116 Forecast of SV availability in Siebeldingen on September 16, 2014, 18:00-06:00.

Trial period 20:00-02:00 with PlanningOnline (http://www.trimble.com/GNSS-PlanningOnline/#/NumSats).

Fig. 116 shows the predicted SV availability for September 16 in the test period up to the morning of September 17. Due to the cyclically better availability (8-12 SV) during the night in summer 2014, the tests were primarily carried out during these times. For the following night, the same constellation (data not shown) was predicted about four minutes earlier.

The `ant_down` and `ant_up` variants were examined for possible effects on the relative position accuracy. For this purpose, rows 1 and 2 of plot 23 were each repeated three times with the corresponding antenna setting and the accuracy in the transverse and longitudinal directions was determined based on the position data of the internal RTK system. Some selected data series are shown below.

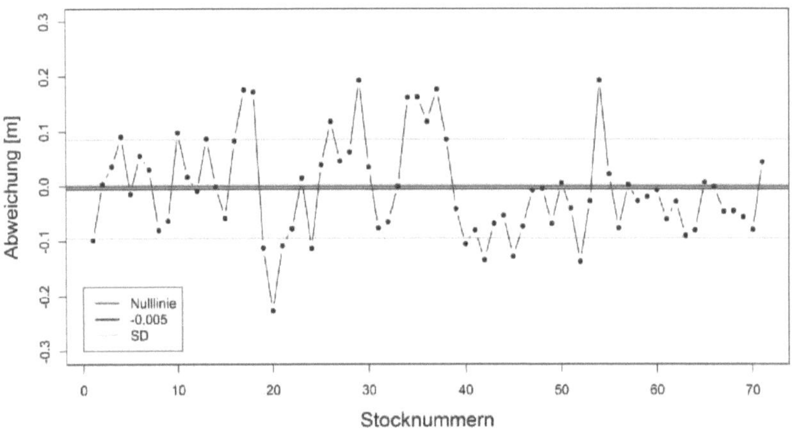

Fig. 117 Relative deviations in transverse direction from the lane in P23, row 1, ant_bottom, August 29, 2014.

Fig. 117 shows the relative deviations as a function of the respective floor position in the `ant_bottom` variant. In this analysis, the mean value was calculated from the actual data, taking into account the sign. The sign provides information about the direction of the deviation in relation to the direction of travel and in relation to the UTM coordinate system.

On average, the actual track (zero line) deviated by -5 mm from the theoretical track (westbound). The scatter of the individual values is quite high, the standard deviation is ± 0.089 m (Table 10). The maximum fluctuation range is 420 mm.

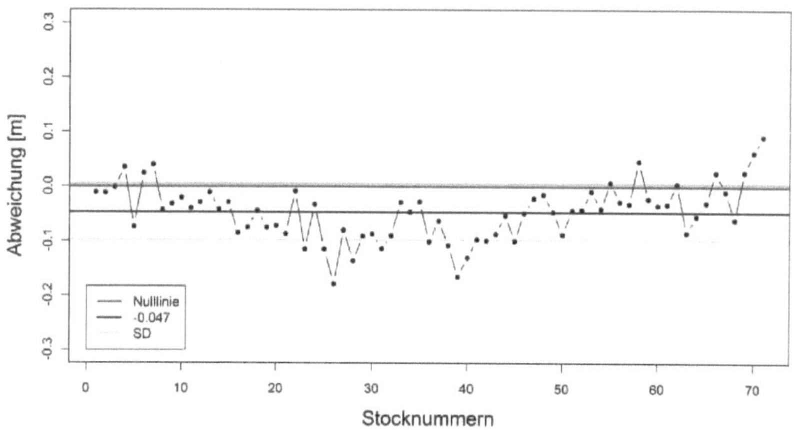

Fig. 118 Relative deviations in transverse direction from the lane in P23, row 1, ant_up, August 29, 2014.

In the `ant_up` setting, the mean transverse deviation from the lane was 47 mm (westbound) (Fig. 118). The scatter was significantly reduced, the value for the standard deviation was reduced by a factor of 2. Similar results were obtained for the scatter of the individual values, which amounted to 270 mm (Tab. 10).

Tab. 10 Descriptive statistics of the relative deviations in transverse and longitudinal direction from the lane in P23, row 1, variants ant_down/ant_up.

	Transversal		***Longitudinal***	
	ant_bottom	ant_above	ant_bottom	ant_above
Mean value (xχ) [m]	-0,005	-0,047	0,1226	0,120
Standard deviation (sx)	0,089	0,051	0,083	0,097
Minimum [m]	-0,227	-0,179	-0.091	-0,188
Maximum [m]	0,193	0,091	0,308	0,301

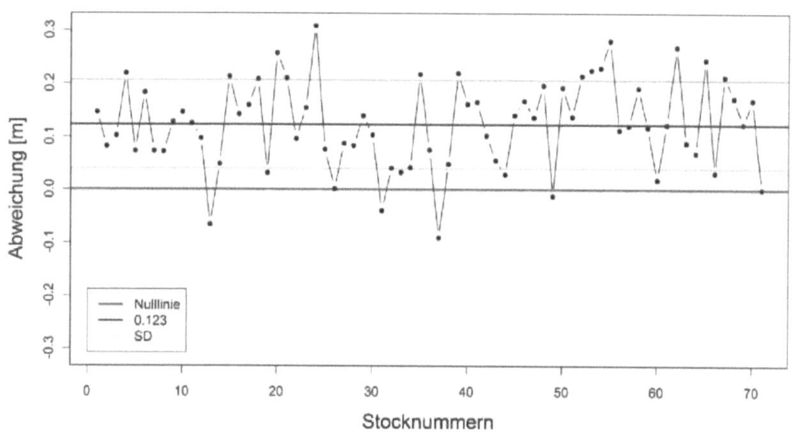

Fig. 119 Relative deviations in longitudinal direction from the lane in P23, row 1, ant_bottom, August 29, 2014.

As expected, the deviations were higher in the longitudinal direction than transverse to the direction of travel. The mean value for the `ant_downward` setting was 123 mm (Fig. 119). This means that the vehicle traveled beyond the stopping point in a northerly direction by this value. The standard deviation was 83 mm and the spread of the individual values was 399 mm.

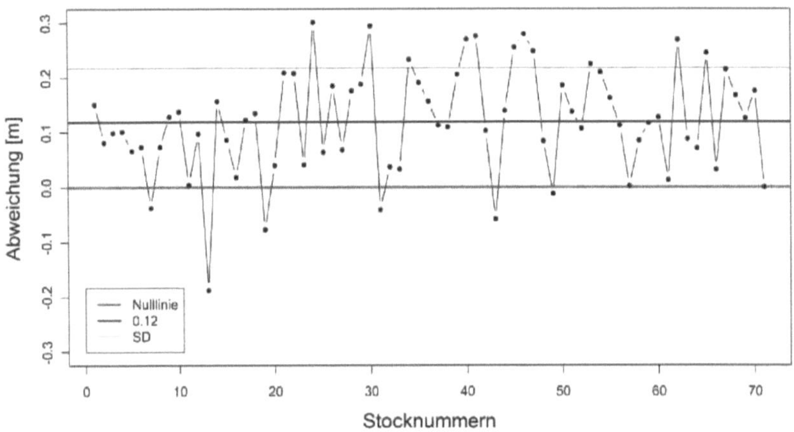

Fig. 120 Relative deviations in longitudinal direction from the lane in P23, row 1, ant_up, August 29, 2014.

Contrary to expectations, the relative deviations in the `ant_up` setting became larger in the longitudinal direction (Fig. 120). Although the mean value was reduced to 120 mm, the dispersion became larger. The standard deviation was 97 mm, the individual value scatter was 489 mm.

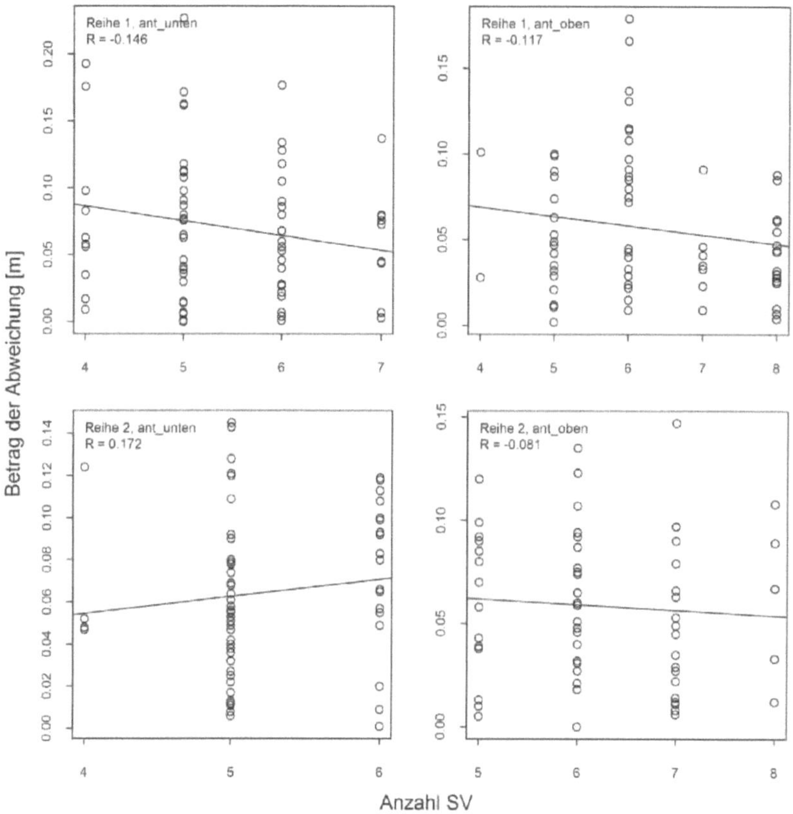

Fig. 121 Amounts of the relative position accuracy as a function of the number of available SV in P23, rows 1 (top) and 2 (bottom), in the ant_down (left) and ant_up (right) variants. Test period: August 29, 2014, 11:30-13:00.

Finally, a possible dependence of the relative positional accuracy on the number of available SVs was investigated. For this purpose, the deviations from the zero line listed above were compared as absolute amounts with the number of SV received (Fig. 121). The regression lines show a weak correlation between decreasing relative deviations with a higher number of available SVs for the variants row 1 `ant_down`, row 1 `ant_up` and row 2 `ant_down`. The negative correlation coefficient according to Spearman confirms this weak correlation. In the variant row 2 `ant_bottom`, on the other hand, there is a weak positive correlation between increasing deviations and a larger number of SVs.

Impact force calculation

Based on the data from the application trial on September 16 and 17, 2014, the application times and thus the impact force were examined. The scoring task consists of:

- Approach to the point (time depends on the route),
- Waiting time to stabilize the vehicle fluctuation (200 ms),
- Position alignment of the MKS (4 500 ms),
- Picture draw,
- Waiting time for response from the photocomputer (up to 6,000 ms),
- Waiting time (500 ms).

This results in a maximum dwell time at the stop position of 11,200 ms without taking into account the travel time between the stops. The evaluation of the log files of some selected rows showed that the response of the image computer took place within 100 ms after a successful image draw. The next job was then started (waiting time 500 ms). If the timeout of 6,000 ms was exceeded, the PHENObotControl application software interrupted job execution until the error was rectified. Before the SMS alert was implemented, the time until further execution depended on the user's reaction time (noticing the error, calling the PHENObot, eliminating the error). According to the evaluation of the logs, the variation in response time ranges from a few seconds to several minutes. In the case of successful

After deducting the waiting time not required, a theoretical dwell time of 5,300 ms per floor remains after the image draw. To determine the actual time for scoring a floor, the time from the end of the waiting time at the position of the previous floor to the start of the approach of the following floor was determined. This is used to determine the required approach time after deducting the theoretical dwell time. The data for row 9 in plot 23 is shown below as an example (Table 11).

Tab. 11 Evaluation of the application times during the scoring in plot 23, row 9 on September 16, 2014.

	P23 Series 9
Minimum [ms]	9 000
Maximum [ms]	15 000
Standard deviation (sx)	1,214
Mean value *(x)* [ms]	11 750

Theoretical dwell time [ms]	5 300
Average arrival time [ms]	6 450

This variation in the required time of 6,000 ms is achieved by shifting the stopping position to the middle of the arch, the variable cane spacing in the case of missing or omitted canes and row openings for crosswise movement. With a total number of 2,726 canes to be scored, a total time of 534 minutes was to be expected just for scoring. Added to this was the time required to approach the next row. About 1 minute was needed for each turning process. This resulted in a total time for driving and scoring plot 23 with 43 rows (42 turns) of 576 minutes, or 9.6 hours (9 hours 36 minutes). With an area to be bonitized of 0.5725 ha, this results in an area output of 0.0596 ha h^{-1}or 16.77 h ha^{-1}based on the conditions in plot 23.

5 Discussion

This chapter discusses and interprets the results and evaluates the validity of the theses formulated at the beginning.

5.1 Carrier vehicle

The concept of the electrically driven tracked vehicle proved to be entirely suitable, taking into account all the results from the design phase and the test drives. The use of tracked undercarriages made it possible to achieve a stable position and smooth driving characteristics of the finished bonus system. The armored steering concept significantly reduced the design and control effort. It was not necessary to produce a steering knuckle, as most of the travel movements were purely linear with only minor angle corrections. With the turning processes at the ends of the rows, the current configuration involves turning on the spot by turning the crawler tracks in the opposite direction. However, the influence on the sward and the natural vegetation in the headland area should be noted. During normal scoring procedures, for example at weekly intervals, only minor effects on the greening can be observed. However, during the test and measurement runs in particular, individual rows and alleys were driven over several times at very short intervals. This resulted in almost complete destruction of the surface vegetation in the turning area of the tracks (Fig. 122).

If wheeled undercarriages were used as an alternative, the low design of the carrier vehicle in this form and the associated smooth running would not be possible. In order to achieve comparable stability, the wheel diameter would have to be correspondingly large, which would also raise the chassis (BLUMENTHAL 1985).

Fig. 122 PHENObot during test run in the vine breeding plot. Significant destruction of the sward in the turning area after several passes.

The construction of the crawler tracks proved to be problematic during testing, especially the two gears. The first defect with total shearing of the tooth flanks was due to a design deficiency. Due to the design, the space in the gearbox was limited. As a result, the final drive gear was thinner than the meshing gear. The contact surface was apparently not sufficient to adequately transmit the large forces at the final drive. As a result, the defect occurred after a test period of around four months. As a result, the gearbox type was replaced by the manufacturer Windegger, which firstly reduced the transmitted force and secondly increased the strength of the gearwheel. A second gearbox defect was caused by an unfortunate impact of the drive wheel on a concrete edge while traveling at too high a speed. This damaged the flanks of the gearwheels in both gearboxes. A further replacement followed. In the course of this and another defect on the control board, the limits of the cooperation with the two companies in South Tyrol became apparent. Due to the large geographical distance, there were understandable delays in the processing of spare parts procurement. Problems with the vehicle libraries could only be discussed and resolved by telephone. Otherwise, the collaboration proved to be very fruitful and successful.

The energy capacity of the batteries proved to be suitable for practical use. However, due to a deep discharge during the test drives, the capacity was reduced towards the end of the project period. After completion of the project, the vehicle owner JKI was therefore recommended to replace the batteries. As part of this purchase, it made sense to increase the total capacity by selecting batteries with a larger individual capacity or by increasing the number of batteries in order to increase the range of the PHE-Nobot without additional charging by the generator (see section 4.8.1).

The design and structure of the navigation system were also appropriate and functional for the applications investigated. However, with reference to the discussion of the measurement results from the test drives, it should be noted at this point that the position accuracy could be significantly increased by compensating for the vehicle inclination. In the course of further development, KMS has also replaced the second GNSS module in the new platforms with an inertial system for detecting orientation (KOFLER 2015). The microcontroller is replaced by an embedded system. In principle, this would make a separate navigation computer superfluous if the embedded system had the appropriate interfaces.

5.2 Camera position control

During the test and project phase, the multi-camera system (MKS) was continuously further developed by the project partner JKI. The test drives provided insights into the design and configuration of the camera and lighting settings. As a result, the required installation space on the platform increased noticeably. This was taken into account in draft III of the camera position control system, which was implemented in practice. Due to the final size of the MKS, it was not possible to maintain the initial goal of accommodating two camera systems on the platform in the future. The choice of aluminum profiles for the flexible construction of the base frame was very advantageous. This meant that deviations from the design and subsequent optimizations, as well as additional attachments, could be easily implemented . In the case of the use of the cut-to-size aluminum sheets from design II, subsequent changes were always associated with additional drilling.

The choice of dimensions, travel paths and swivel angles was confirmed by the findings and results of the test and measurement runs. It was shown that the fourth degree of freedom "swivel axis II" is not necessary for the scoring task to be carried out. The swivel of the FMD around the image axis would even result in a disadvantageous effect. In conventional trellis systems, the foliage wall of the vine row is always aligned parallel to the ground, regardless of the slope inclination. When moving and zeroing the camera position control, the FMD is guided parallel to the platform and therefore parallel to the ground. However, if a position-dependent levelling of the FMD in swivel axis II were to be carried out, the image section could shift with increasing slope inclination and thus lead to a loss of image information in the edge and corner areas. For this reason, leveling around pan axis II was not used. A positive side effect can also be seen in the energy saved by not positioning by the fourth degree of freedom.

The automatic leveling of the swivel axis I according to the data from the tilt sensor worked very reliably during the tests. A correction of the height setting at large swivel angles was not necessary according to the evaluations of the image data. There was no loss of image information due to cut-off foliage wall areas. The

Setting the horizontal distance to the foliage wall and the pick-up height are preselected manually via the control pad according to the current configuration. Automation can be implemented with little programming effort. Additional distance and height information would have to be implemented in the coordinate database and further processed by the transformation scripts. In the medium term, however, it is planned to retain the manual solution, as the settings only need to be made once for each plot to be scored.

5.3 PHENObot

The modular design with a platform construction and detachable cam position control system proved its worth during the test phase. The entire superstructure can be separated from the carrier platform within a few minutes, especially for maintenance and repair work. All that is needed is an Allen key and a forklift truck or pallet truck. During the repair phase of the second gearbox defect, it was possible to continue working on and optimizing the remote position control system (Fig. 53 and Fig. 54).

The operability of the entire system via the two operating areas was rated positively by all users. All configuration areas can be operated separately from each other. For example, no user input is required via the computer to manually adjust the position of the MKS. All entries can be made via clearly labeled switching elements on the control cabinet.

The division into two sub-projects resulted in the use of two separate computers with different operating systems. To optimize the system layout, porting to one computer would be conceivable here, also taking into account an embedded system on the vehicle platform. During the test phase, however, this was not feasible due to the spatial distance between the two project groups.

The high-quality stepper motor and control technology used has proven to be extremely robust, especially in bad weather. As a result, inspections can also be carried out during periods of bad weather. The use of crawler tracks also has a positive effect here (BLUMENTHAL 1985).

5.4 Software development

A significant part of the project phase was used to develop various software and control applications. An intensive evaluation of the required components was carried out during testing. The development proceeded in several stages. On the one hand, the behaviour of the carrier vehicle with basic functions of a development variant of PHENO- botControl was analyzed and optimized. Secondly, the camera position control system was developed in parallel, using the control algorithms supplied by the manufacturer and implementing additional functions, such as automatic levelling with additional sensors. The last central component of the overall system was the development and refinement of the task processing.

The PHENObotControl application development area in particular had a steep learning curve, including the implementation of a number of advanced functions that go well beyond the requirements of the basic functions. These included, for example, logging and alarm functions. The interaction of the navigation and position control functional areas could also be easily implemented.

The structure of a stock information database was created based on the experience gained from the first preliminary tests. The workflow for generating the job and project files was based on this. The findings from the early test phase led to the intensive use of the GIS applications GRASS GIS and QGIS, whose respective advantages (BILL 2010, STEINIGER & HAY 2009) were specifically combined. The use of the bash-

Scripts for controlling and processing the transformation and the job file output proved to be flexible and still efficient in the present scope. If the requirements increase and the work orders become more complex, it might be conceivable to create a separate application in a higher-level language, e.g. C++ using the Qt IDE.

5.5 Evaluation of the test phase

The findings of the test drives, especially those on the test area in the courtyard of the Institute of Technology, show the need for reliable availability of satellites during automatic driving operation. The lower threshold of at least five SV should, in principle, ensure sufficiently accurate operation. In some cases, especially under critical conditions with regard to availability, inaccuracies and unpredictable reactions of the carrier platform occurred at times. For example, uncontrolled rotation around the vehicle axis occurred, particularly in the threshold area, when a new driving task was initialized, which was attributed to incorrect parameters in the vehicle library. This error should be eliminated in the course of a firmware update of the carrier platform. After this, the behavior no longer occurred until the project was completed. Unfortunately, the problem of correction signal loss with very high SV availability could not be solved causally. The cause is suspected to be either a fault in the firmware of the GNSS module of the base station or the application based on it. These sub-areas are subject to the responsibility and interests of the supplying companies and are therefore protected. It was not possible to analyze the source code for troubleshooting purposes. Contacting and attempting to solve the problem with the module manufacturer was unsuccessful. As a workaround, a variable elevation mask was implemented in the base station software. If an availability of more than nine SV occurs, the mask is increased until the threshold is no longer exceeded. This solution proved to be practicable in the course of further application. No negative effects on the accuracy or stability of the correction signal could be determined, taking into account the general definition of location accuracy according to DODEL (2010).

The calibrations of the RTK systems for datum transformation from WGS84 to ETRS89 (BECKER & HEHL 2012) proved to be practical and sufficiently accurate.

A possible influence of the fully developed foliage wall on the reception strength was derived from the experience with poor availability on the test area technology.

At times of full foliage, the antenna unit is in the initial situation at about half the height of the foliage wall. This results in elevation angles >45° and therefore inevitable shadowing and multipath effects on the foliage. For this reason, a height adjustment was designed for the antenna unit, which brings both GNSS antennas into the area of the summit zone. This results in elevations <15° depending on the height of the foliage. The tests carried out in Geisenheim and Siebeldingen with different height settings consistently show a positive effect on availability at the top setting `ant_up`. In addition, there was a reduction in scattering in the transverse direction with the `ant_up` setting. However, the higher position of the GNSS antenna increases the influence of the landing position on the absolute positioning. In principle, all GNSS systems assume the position vertically below the antenna. Without compensation for the vehicle inclination, this shifts on the basis of trigonometry when the vertical axis is tilted. In the PHENObot's navigation system, tilt compensation is not provided in the current expansion stage. This results in the lane shifts in the transverse direction due to the moderate side slope in plot 23 (Fig. 118). The same applies to the deviations in the longitudinal direction, whereby this error is also dependent on the travel movement and direction of travel. Along the

In the direction of travel, the positioning is dependent on the dynamics of the driving speed. In all evaluations, the vehicle drove beyond the theoretical stopping point and only stopped afterwards. This is consistent with the results of NORREMARK et al. (2007). In addition, the higher antenna position results in a greater influence due to rolling movements of the vehicle when stopping. A bouncing of the platform when reaching the point and stopping was visually recognizable. This behavior was favored by unfavorable terrain characteristics (e.g. road elevations).

In order to investigate whether the improved availability of `ant_up` positioning had a positive effect on the position accuracy, the amounts were set in relation to the number of SVs received. Only a weak correlation was found, although there was a tendency towards an improvement in accuracy (Fig. 121).

The absolute accuracy during the repeated run was within the range of values known from the literature (GAN-MOR et al. 2007, NORREMARK et al. 2007, SUN et al. 2010, PEREZ-RUIZ et al. 2012). The influence of the lack of slope compensation must be taken into account. NORREMARK et al. (2007) found up to 59 times greater position errors when tilt compensation was omitted. It was found that the absolute error in the transverse direction is about twice as large as the accuracy specified by the manufacturer of the GNSS modules. This finding is also consistent with the above-mentioned investigations. The discrepancy is due to sources of error in GNSS-based positioning. It could not be definitively clarified whether the RTK system used was solely responsible for the error. In one of the two series of measurements investigated, no significant mean value difference could be found between the navigation system and the reference system. In the case of the second pair of mean values, whose difference was statistically significant, the error was two times greater compared to the first pair of mean values. The role of the lack of slope compensation should be emphasized again here.

In order to finally investigate the absolute accuracy, a reference measurement with a tachymetric method would be conceivable, the accuracy of which is ten times greater. For time and financial reasons, this method could not be used in the context of this work. In the author's opinion, the absolute accuracy found during the measurement and test runs is quite sufficient for the requirements of the scoring task. For this reason, the relative position error and its fluctuation were also investigated. No significant deviation was found between the rows driven over during the four replicates. The position error (38 mm transversal; 73 or 74 mm longitudinal) (Fig. 106) is again within the range known from the literature. Taking into account the scoring requirements, a relative accuracy of <50 mm is desirable. According to this definition, the transversal error found is positive, the longitudinal deviation rather negative. Nevertheless, measured by the borehole results, the task was satisfactorily fulfilled by the image data, as the variability of the vines (stopping position in the middle of the bending vine) exceeds this longitudinal position error by a factor of about two.

The cane coordinates used were recorded in advance after planting the trial plots using an RTK measurement system (see Fig. 29). The vertical alignment of the GNSS antennas above the planting position must be observed. Especially during the vegetation phase, this was not always possible due to the developed foliage wall. This resulted in shifts in the transverse and longitudinal direction from the actual location of the vine. These discrepancies became noticeable during the first test runs due to oscillating and wobbling movements of the PHENObot, especially in the transverse direction. To eliminate this, a smoothing function was implemented in the transformation script for outputting the job files. Comparative runs in a row before and after smoothing showed a clearly perceptible, smoother driving behavior without major corrective movements. It is not only the energy savings that should be emphasized here due to the overall shorter

The new system not only enables more precise travel paths and corrective movements, but also more precise alignment of the FMD virtually perpendicular to the foliage wall.

The analysis of the data to evaluate the potential impact and performance of the scoring system yielded very positive results. Including all travel and turning times, it took around 580 minutes to score plot 23. With a number of 2,726 canes, this results in an average scoring time of 12.8 seconds per cane. ROSCHER et al. (2014) required around four minutes per plant for manual scoring of the bee diameters using a caliper gauge (ROSCHER et al. 2014). This reduces the time required to record the typical *yield* characteristic by a factor of almost 19. With both methods, the subsequent data processing must also be taken into account. The automatic image recording allows the data to be used for the investigation of other parameters, such as *disease infestation*. No additional scoring of notes is required. Furthermore, automated scoring is not subject to any subjective influences and can be carried out independently of human operator fatigue. It is only necessary to ensure that the platform is sufficiently supplied with energy. A further acceleration of the scoring could be achieved by capturing images while driving without stopping. At present, however, it is still necessary to remain stationary to ensure optimum image quality.

5.6 Assessment of the theses

The theses formulated at the beginning can be assessed as follows after evaluating the results, taking into account the current state of knowledge:

I. Taking into account the defined task requirements, the PHENObot has succeeded in constructing an automatically moving platform that can be used without restriction in the large-scale systems of a vineyard.

II. The accuracy achieved was compared with literature data and is below the defined threshold when considering the absolute sizes in the transverse direction. Although higher errors occurred along the direction of travel, these were not detrimental in practice when considering the general variability of the vineyards.

III. The automated processing of the existing stock coordinates was successfully implemented. The receipt system can read and process the exported order lists directly and immediately. If the range of functions is extended, porting to a more complex software environment is conceivable.

IV. The multi-camera system with integrated lighting unit was implemented in a camera position control system with four (currently used: three) possible degrees of freedom. The height and distance adjustment is carried out manually according to user requirements. Existing transverse terrain inclinations are adaptively compensated before each image acquisition in order to be able to create a parallel alignment of the image plane to the foliage wall plane.

V. The performance of the scoring system considerably exceeds the potential of manual phanotyping. The required level of 250 vines per hour can be surpassed even when the travel and turning times are taken into account. The data acquisition is completely objective without any influence of human fatigue.

5.7 Outlook

In its current configuration, the bonus system (Fig. 123) could be used not only in grapevine breeding, but also in yield-oriented viticulture. It is conceivable that it could be used in large grape-producing companies or cooperatives, as a component of the quality-oriented payment system. The cultivation of new and additional sensors for data extraction is also possible.

The satellite-controlled carrier vehicle can also be used with other superstructures for maintenance work in viticulture, as well as other special and space crops. The drive and control concept could also be transferred to new self-propelled and automatic or autonomous applications.

Fig. 123 PHENObot in the final version during the preliminary tests for the application trial.

6 Summary

The cultivation of vines can look back on a long tradition, which at the same time is characterized by constant further development. Great importance is attached to fungal resistance when breeding new grape varieties. Increasingly, methods of precision agriculture are being adapted and used in the management of vineyards. The PHENOvines research project attempted to link these two areas with each other. The automated, self-propelled PHENObot scoring platform was developed to accelerate and objectify the phanotyping required during vine breeding.

This book focuses on the conceptual and constructive work to create this scoring platform, its navigation and the management of the sensor system for image data acquisition. Another component is the experimental testing and trial phase. At the beginning of the project, drive concepts were evaluated and finally a large crate transporter already on the market, operated with electric belt drives, was selected as the carrier vehicle. The automatic track guidance along floor coordinates recorded using NAVSTAR GPS was ensured by equipping it with an RTK navigation system. The position data alone is used for navigation. In addition, ultrasonic sensors and a mechanical collision shield for obstacle detection are used as safety devices. The scoring system is operated via the specially developed PHENObotControl 1.0 control application. This requires text-based job files, which are generated using the cane coordinates by means of script-based transformation and processing in the GIS application GRASS GIS. The scoring process is divided into approaching the stopping point directly in front of the vine, levelling the multi-camera system (MKS), drawing images, saving images and moving on to the next vine. Leveling is carried out using the data from a tilt sensor in a position control system with four possible degrees of freedom. When the specified position is reached, the navigation system sends a release command with position and identification data to the image acquisition system, which contains five cameras of different wavelength ranges as well as an LED lighting system. In order to guarantee the objectivity and consistent image quality, the scoring takes place primarily in the dark or under neutral lighting conditions. The energy for the electric drive is stored in a battery pack, which can also be recharged via a generator based on the hybrid drive principle.

During the test phase, all functions of the PHENObot were tested and further developed. Measurement runs were used to determine and quantify the position errors that occurred and their sources. On average, the transversal position deviations transverse to the direction of travel were less than 50 mm. This is sufficiently accurate to be able to optimally align the FMD for image acquisition in front of a vine. An application test in two consecutive nights concluded the test phase at the Siebeldingen site. Image information was collected from 2,726 vines at a scoring rate of a good 280 vines per hour or just under 600 $m^{(2)} h^{(-)}{}_{1}$. Compared to manual scoring of the yield parameter, the performance was increased by a factor of almost 19. In addition, the recorded images provide further information for assessing other parameters. In the additional time available, the user can devote himself to the evaluation and interpretation of the obtained phenotypic data. In this way, a contribution can be made to reducing the breeding time of a new grape variety.

7 Bibliography

Monographs and articles

ADAM, J., AUGATH, W., BOUCHER, C., BRUYNINX, C., DUNKLEY, P., GUBLER, E., GURTNER, W., HORNIK, H., MAREL, H., SCHLÜTER, W. (2000): The European reference system coming of age, Geodesy Beyond 2000, Springer, pp. 47-54.

ADRIAN, A., NORWOOD, S., MASK, P. (2005): Producers' perceptions and attitudes toward precision agriculture technologies. Computers and Electronics in Agriculture 48, (3), pp. 256-271.

ALBERTZ, J. (2009): Einführung in die Fernerkundung - Grundlagen der Interpretation von Luft- und Satellitenbildern, Wiss. Buchges., Darmstadt. 4th ed.

AQEEL-UR-REHMAN, ABBASI, A., ISLAM, N., SHAIKH, Z. (2014): A review of wireless sensors and networks' applications in agriculture. Computer Standards & Interfaces 36, (2), pp. 263-270.

ARNÓ, J., CASASNOVAS, J., DASI, M., ROSELL, JR (2009): Review. Precision viticulture. Research topics, challenges and opportunities in site-specific vineyard management. Spanish Journal of Agricultural Research 7, (4), pp. 779-790.

AUERNHAMMER, H. (1994): Introduction. Computers and Electronics in Agriculture 11, (1). p. 1.

AUERNHAMMER, H. (2000): The Agricultural BUS System LBS and the Satellite Positioning System GPS as Key Technologies for Process Control in Crop Production - Papers of the 21st GIL Annual Conference in Freising-Weihenstephan 2000 (Volume 13). GESELLSCHAFT FÜR INFORMATIK IN DER LAND-, FORST- UND ERNAHRUNGSWIRTSCHAFT, ED.

AUERNHAMMER, H. (2001): Precision farming - the environmental challenge. Computers and Electronics in Agriculture 30, (1-3), pp. 31-43.

AUERNHAMMER, H. (2002): Automatic farm data collection in arable farming and its application. Arable farming of the future, 45.

AUERNHAMMER, H., DEMMEL, M., MUHR, T., ROTTMEIER, J., WILD, K. (1994): GPs for yield mapping on combines. Computers and Electronics in Agriculture 11, (1), pp. 53-68.

AUERNHAMMER, H., DEMMEL, M., SPANGLER, J. (1999): Farm data documentation with LBs and GPs for tractor-implement combinations - Tagung Landtechnik.

AUTOVIMATION, in collaboration with RÜGER, P. (2014): 3D-CAD design of the multi-camera system II. Personal communication.

BANGERT, W., KIELHORN, A., RAHE, F., ALBERT, A., BIBER, P., GRZONKA, S., HAUG, S., MICHAELS, A., MENTRUP, D., HÄNSEL, M. (2013): Field-Robot-Based Agriculture: RemoteFarming. 1 and BoniRob-Apps. VDI Reports, (2193), pp. 439-446.

BASTIAANSSEN, W., MOLDEN, D., MAKIN, I. (2000): Remote sensing for irrigated agriculture: examples from research and possible applications. Agricultural Water Management 46, (2), pp. 137-155.

BATTE, M., EHSANI, M. (2006): The economics of precision guidance with auto-boom control for farmer-owned agricultural sprayers. Computers and Electronics in Agriculture 53, (1), pp. 28-44.

BAUER, M. (2011): Vermessung und Ortung mit Satelliten - Globale Navigationssatellitensysteme (GNSS) und andere satellitengestützte Navigationssysteme, Wichmann, Berlin [u.a.]. 6th ed.

BBA (2001): Developmental stages of mono- and dicotyledonous plants - BBCH monograph.

BECKER, M., HEHL, K. (2012): Geodasie, WBG (Wissenschaftliche Buchgesellschaft), [p.l.].

BIBER, P., WEISS, U., DORNA, M., ALBERT, A., eds. (2012): Navigation system of the autonomous agricultural robot Bonirob.

BILL, R. (2010): Grundlagen der Geoinformationssysteme, Wichmann, Berlin.

BLACKMORE, B., FOUNTAS, S., GEMTOS, T., GRIEPENTROG, H. (2009): A Specification for an Autonomous Crop Production Mechanization System. Acta Hort. (ISHS), (824), pp. 201-216.

BLOSS, R. (2014): Robot innovation brings to agriculture efficiency, safety, labor savings and accuracy by plowing, milking, harvesting, crop tending/picking and monitoring. Industrial Robot 41, (6), pp. 493-499.

BLUMENTHAL, R. (1985): Technisches Handbuch Traktoren, VEB Verlag Technik.

BOGUE, R. (2013): Can robots help to feed the world? Industrial Robot 40, (1), pp. 49.

BØHRNSEN, A. (2009): Drive with lane indicator or steer automatically. Profi, (7), pp. 74-77.

BOTTINGER, S. (2010a): Information technology for use on agricultural machinery - papers presented at the 30th GIL Annual Conference in Hohenheim 2010 - Precision Agriculture Reloaded - information-supported agriculture. GESELLSCHAFT FÜR INFORMATIK IN DER LAND-, FORST- UND ERNAHRUNGSWIRTSCHAFT, ED.

BÖTTINGER, S. (2010b): Robots and automation for machine management. KTBL, ed.

BRUNNER, D., in collaboration with RÜGER, P. (2014): Geoidundulation for the Geisenheim area. Personal communication.

BVV (2009): UTM mapping and UTM coordinates. STATE OFFICE FOR DIGITIZATION, BROADBAND AND SURVEYING, BAVARIA, ED.

CALDEWEYHER, D., ZHANG, J., PHAM, B. (2006): OpenCIS-Open Source GIS-based web community information system. International Journal of Geographical Information Science 20, (8), pp. 885-898.

CHEN, D., SHAMS, S., CARMONA-MORENO, C., LEONE, A. (2010): Assessment of open source GIS software for water resources management in developing countries. Journal of Hydro-environment Research 4, (3), pp. 253-264.

CIEPLUCH, B., JACOB, R., MOONEY, P., WINSTANLEY, A., eds (2010): Comparison of the accuracy of OpenStreetMap for Ireland with Google Maps and Bing Maps, University of Leicester.

COX, S. (2002): Information technology: the global key to precision agriculture and sustainability. Computers and Electronics in Agriculture 36, (2-3), pp. 93-111.

CUBERO, S., ALEIXOS, N., ALBERT, F., TORREGROSA, A., ORTIZ, C., GARCÌA- NAVARRETE, O., BLASCO, J. (2014): Optimized computer vision system for automatic pre-grading of citrus fruit in the field using a mobile platform. Precision Agric 15, (1), pp. 80-94.

DE-AN, Z., JIDONG, L., WEI, J., YING, Z., YU, C. (2011): Design and control of an apple harvesting robot. Biosystems Engineering 110, (2), pp. 112-122.

DEERY, D., JIMENEZ-BERNI, J., JONES, H., SIRAULT, X., FURBANK, R. (2014): Proximal Remote sensing Buggies and Potential Applications for Field-Based Phenotyping. Agronomy 4, (3), pp. 349-379.

DELCOURT, H., BAERDEMAEKER, J. DE (1994): soil nutrient mapping implications using GPs. Computers and Electronics in Agriculture 11, (1), pp. 37-51.

DiN (2007): DiN EN iso 19111:2007-10 - Geoinformation - Coordinate reference systems (iso 19111:2007), Beuth.

DODEL, H., HAUPLER, D. (2010): Satellitennavigation, Springer-Verlag Berlin Heidelberg, Berlin, Heidelberg, 1 online resource (online-Ressource.).

EHRL, M., AUERNHAMMER, H. (2007): X-by-wire via iSoBuS communication network. Agricultural Engineering international: CiGR Journal.

EMMI, L., PAREDES-MADRID, L., RIBEIRO, A., PAJARES, G., GONZALEZ-DE-SANTOS, P. (2013): Fleets of robots for precision agriculture: a simulation environment. industrial Robot 40, (1), pp. 41-58.

European union (2012): Primary and secondary descriptors used in Genres- Projects (Genres081/GrapeGen06).

Fernandez, R., Montes, H., Salinas, C., Sarria, J., Armada, M. (2013): Combination of RGB and multispectral imagery for discrimination of cabernet sauvignon grapevine elements. sensors (Basel, switzerland) 13, (6), pp. 78387859.

Fritz, s., McCallum, i., Schill, C., Perger, C., See, L., Schepaschenko, D., van der Velde, Marijn, Kraxner, F., Obersteiner, M. (2012): Geo-Wiki: An online platform for improving global land cover. Environmental Modeling & software 31, pp. 110-123.

Gan-Mor, S., Clark, R., Upchurch, B. (2007): Implement lateral position accuracy under RTK-GPs tractor guidance. Computers and Electronics in Agriculture 59, (1-2), pp. 31-38.

GEO-KONZEPT (2014): Demonstration geo-konzept Octocopter. Presentation, Geisenheim.

Godfray, C., Beddington, J., Crute, I., Haddad, L., Lawrence, D., Muir, J., Pretty, J., Robinson, S., Thomas, S., Toulmin, C. (2010): Food security: the challenge of feeding 9 billion people. Science (New York, N.Y.) 327, (5967), pp. 812-818.

Goodchild, M. (1992): Geographical data modeling. Computers & Geosciences 18, (4), pp. 401-408.

Gorres, B., Mayer, M., Nothnagel, A., Heck, B. (2012): Reference systems for GNSS - DVW-Merkblatt 5-2012. DVW - Gesellschaft für Geodasie, Geoinformation und Landmanagement e.V., ed.

Griepentrog, H. (2010): Automation in the outdoor economy. KTBL, ed.

Gruber, F., Joeckel, R. (2014): Formelsammlung für das Vermessungswesen, springer Fachmedien Wiesbaden GmbH, Wiesbaden. 17th ed.

Hall, A., Lamb, D., Holzapfel, B., Louis, J. (2002): Optical remote sensing applications in viticulture-a review. Australian Journal of Grape and Wine Research 8, (1), pp. 36-47.

Hall, A., Louis, J., Lamb, D. (2003): characterizing and mapping vineyard canopy using high-spatial-resolution aerial multispectral images. Computers & Geosciences 29, (7), pp. 813-822.

Harzer, c. (2014): Gis-Report 2014/15.

Hayashi, S., Shigematsu, K., Yamamoto, S., Kobayashi, K., Kohno, Y., Kamata, J., Kurita, M. (2010): Evaluation of a strawberry-harvesting robot in a field test. Biosystems Engineering 105, (2), pp. 160-171.

Hemisphere (2012): Eclipseii_Data_sheet_WEB_2.2012 - Data sheet.

JI, W., ZHAO, D., CHENG, F., XU, B., ZHANG, Y., WANG, J. (2012): Automatic recognition vision system guided for apple harvesting robot. Computers & Electrical Engineering 38, (5), pp. 1186-1195.

JOHNSON, L., BOSCH, D., WILLIAMS, D., LOBITZ, B. (2001): Remote sensing of vineyard management zones: implications for wine quality. Applied Engineering in Agriculture 17, (4), pp. 557-560.

Kadisch, E., Müller, E., Lipps, H.-P. (2008): Weinbau - Der Winzer, Ulmer, Stuttgart.

KEICHER, R. (2002): Stock-oriented automatic tracking of agricultural machinery in row crops with the aid of digital image processing. Dissertation INSTITUT FÜR LANDTECHNIK GIEBEN, JUSTUS LIEBIG UNIVERSITY, ED.

KEICHER, R., SEUFERT, H. (2000): Automatic guidance for agricultural vehicles in Europe. Computers and Electronics in Agriculture 25, (1-2), pp. 169-194.

KICHERER, A., in collaboration with RÜGER, P. (2013a): Evaluation results image acquisition with multi-camera system without artificial lighting. Personal communication.

KICHERER, A., in collaboration with RÜGER, P. (2013b): Visual material for the PHENOvines research project. Personal communication.

KICHERER, A., HERZOG, K., PFLANZ, M., WIELAND, M., RÜGER, P., KECKE, S., KUHLMANN, H., TOPFER, R. (2015): An Automated Field Phenotyping Pipeline for Application in Grapevine Research. Sensors 15, (3), pp. 4823-4836.

KICHERER, A., ROSCHER, R., HERZOG, K., SIMON, S., FORSTNER, W., TOPFER, R. (2013): BAT (Berry Analysis Tool): a high-throughput image interpretation tool to acquire the number, diameter, and volume of grapevine berries. Vitis 52, (3), pp. 129-135.

KOFLER, A., in collaboration with RÜGER, P. (2015): Functionality of vehicle control and library. Personal communication.

KONDO, N., MONTA, M., FUJIURA, T. (1996): Fruit harvesting robots in Japan. Advances in space Research 18, (1-2), pp. 181-184.

KONRAD, H., in collaboration with RÜGER, P. (2015): information on the experimental vineyard for grapevine breeding at Geisenheim University. Personal communication.

KRÜGER, G., SPRINGER, R., LECHNER, W. (1994): Global Navigation satellite systems (GNss). Computers and Electronics in Agriculture 11, (1), pp. 3-21.

KRUIZE, J., ROBBEMOND, R., SCHOLTEN, H., WOLFERT, J., BEULENS, A. (2013): improving arable farm enterprise integration - Review of existing technologies and practices from a farmer's perspective. Computers and Electronics in Agriculture 96, pp. 75-89.

KTBL (2013): Weinbau und Kellerwirtschaft, KTBL, Darmstadt. 15th ed.

KÜBLER, H. (2010): Perspectives for farm management information systems. KTBL, ed.

LAMB, D., BRAMLEY, R., HALL, A. (2002): Precision viticulture-an Australian perspective. XXVi international Horticultural congress: Viticulture-Living with Limitations 640, pp. 15-25.

LAMB, D., FRAZIER, P., ADAMS, P. (2008): Improving pathways to adoption: Putting the right P's in precision agriculture. Computers and Electronics in Agriculture 61, (1), pp. 4-9.

LAMB, D., WEEDON, M., BRAMLEY, R. (2004): Using remote sensing to predict grape phenolics and colour at harvest in a cabernet sauvignon vineyard: Timing observations against vine phenology and optimizing image resolution. Australian Journal of Grape and Wine Research 10, (1), pp. 46-54.

LARSEN, W., NIELSEN, G., TYLER, D. (1994): Precision navigation with GPs. Computers and Electronics in Agriculture 11, (1), pp. 85-95.

LECHNER, W., BAUMANN, S. (2000): Global navigation satellite systems. Computers and Electronics in Agriculture 25, (1-2), pp. 67-85.

LEE, W., ALCHANATIS, V., YANG, c., HIRAFUJI, M., MOSHOU, D., LI, c. (2010): sensing technologies for precision specialty crop production. Computers and Electronics in Agriculture 74, (1), pp. 2-33.

LENZ, T., OTTO, M. (2013): Praxisratgeber Maschinensicherheit: Fallbezogene Handlungsanleitungen nach der neuen Maschinenrichtlinie, Forum Verlag Herkert, Merching.

LINZ, A., BRUNNER, D., FEHRMAN, J., GRIMSEL, M., HERLITZIUS, T., KEICHER, R., RUCKELSHAUSEN, A., SCHWARZ, H.-P., WUNDER, E. (2014a): "elWObot"- A DieselElectric Autonomous Sensor Controlled Service-Robot for Orchards and Vineyards. 4th Conference on Machine Control & Guidance, March 19-20, (2014) Proceedings, pp. 16-23.

LINZ, A., RUCKELSHAUSEN, A., WUNDER, E., HERTZBERG, J. (2014b): Autonomous Service Robots for Orchards and Vineyards: 3D Simulation Environment of Multi Sensor-Based Navigation and Applications.

LONGO, D., MUSCATO, G. (2013): Design and Simulation of Two Robotic Systems for Automatic Artichoke Harvesting. Robotics 2, (4), pp. 217-230.

LONGO, D., MUSCATO, G., CARUSO, L., CONTI, A., SCHILLACI, G., eds (2012): Design of a remotely operable sprayer for precision farming application.

LONGO, D., PENNISI, A., BONSIGNORE, R., MUSCATO, G., SCHILLACI, G., eds (2010): A Multifunctional Tracked Vehicle Able to Operate in Vineyards Using GPS and Laser Range-finder

Technology.

MANSFELD, W. (2010): Satellitenortung und Navigation - Grundlagen, Wirkungsweise und Anwendung globaler Satellitennavigationssysteme ; mit 65 Tabellen, Vieweg + Teubner, Wiesbaden. 3rd ed.

MAOHUA, W. (2001): Possible adoption of precision agriculture for developing countries at the threshold of the new millennium. Computers and Electronics in Agriculture 30, (1-3), pp. 45-50.

MIELE, A., FLORES, C., FILIPPINI ALBA, JOSE MARIA, AURAND, J.-M. (2014): Precision viticulture in Brazil: Current research status on wine grape. BIO Web of Conferences 3, 01013.

MITASOVA, H., MITAS, L., HARMON, R. (2005): Simultaneous Spline Approximation and Topographic Analysis for Lidar Elevation Data in Open-Source GIS. IEEE Geosci. Remote Sensing Lett. 2, (4), pp. 375-379.

MOSER, M., MÜLLER, G., SCHLEMMER, H. (2012): Handbuch Ingenieurgeodasie, Wichmann, Heidelberg. 4th ed.

NEMENYI, M., MESTERHAZI, P., PECZE, Z., STEPAN, Z. (2003): The role of GIS and GPS in precision farming. Computers and Electronics in Agriculture 40, (1-3), pp. 45-55.

NETELER, M., BOWMAN, M., LANDA, M., METZ, M. (2012): GRASS GIS: A multipurpose open source GIS. Environmental Modeling & Software 31, pp. 124-130.

NICOLAÏ, B., BEULLENS, K., BOBELYN, E., PEIRS, A., SAEYS, W., THERON, K., LAMMERTYN, J. (2007): Nondestructive measurement of fruit and vegetable quality by means of NIR spectroscopy: A review. Postharvest Biology and Technology 46, (2), pp. 99-118.

NIKKILA, R., SEILONEN, I., KOSKINEN, K. (2010): Software architecture for farm management information systems in precision agriculture. Computers and Electronics in Agriculture 70, (2), pp. 328-336.

NOACK, P. O. (2006): Development of lane-based algorithms for the correction of yield data in precision farming. Dissertation. DEPARTMENT OF BIOGENIC RESOURCES AND LAND USE TECHNOLOGY, DIVISION OF TECHNOLOGY IN CROP PRODUCTION, TU MUNICH-WEIHENSTEPHAN, ED.

NOACK, P. O., GEIMER, M., EHRL, M., GRANDL, L., eds. (2010): Virtual coupling of vehicles: Electronic drawbar for agricultural machinery.

NORREMARK, M., **SOGAARD**, H., GRIEPENTROG, H., NIELSEN, H. (2007): instrumentation and method for high accuracy geo-referencing of sugar beet plants. Computers and Electronics in Agriculture 56, (2), pp. 130-146.

PEREZ-RUIZ, M., SLAUGHTER, D., GLIEVER, C., UPADHYAYA, S. (2012): Tractor-based Real-time

Kinematic-Global Positioning System (RTK-GPS) guidance system for geospatial mapping of row crop transplant. Biosystems Engineering 111, (1), pp. 64-71.

PHYTRON (2014): phyMoTioN - Modular multi-axis controller for stepper motors - Product information.

REICHARDT, M., JÜRGENS, C. (2009): Adoption and future perspective of precision farming in Germany: results of several surveys among different agricultural target groups. Precision Agric 10, (1), pp. 73-94.

ROSCHER, R., HERZOG, K., KUNKEL, A., KICHERER, A., TOPFER, R., FORSTNER, W. (2014): Automated image analysis framework for high-throughput determination of grapevine berry sizes using conditional random fields. Computers and Electronics in Agriculture 100, pp. 148-158.

ROTHMUND, M., WODOK, M. (2010): isoBUs-Eine systematische Betrachtung der Norm iso 11783 - Referate der 30. GiL-Jahrestagung in Hohenheim 2010 - precision Agriculture Reloaded - informationsgestützte Landwirtschaft.

GESELLSCHAFT FÜR INFORMATIK IN DER LAND-, FORST- UND ERNÄHRUNGSWIRTSCHAFT, ED.

RUCKELSHAUSEN, A. (2010): Autonomous field robots. KTBL, ed.

RUCKELSHAUSEN, A., BIBER, p., DORNA, M., GREMMES, H., KLOSE, R., LINZ, A., RAHE, F., RESCH, R., THIEL, M., TRAUTZ, D. (2009): BoniRob-an autonomous field robot platform for individual plant phenotyping. precision agriculture 9, (841), p. 1.

RUIZ-ALTISENT, M., RUIZ-GARCIA, L., MOREDA, G., LU, R., HERNANDEZ-SANCHEZ, N., CORREA, E., DIEZMA, B., NICOLAÏ, B., GARCIA-RAMOS, J. (2010): Sensors for product characterization and quality of specialty crops-A review. Computers and Electronics in Agriculture 74, (2), pp. 176-194.

SANKARAN, S., MISHRA, A., EHSANI, R., DAVIS, C. (2010): A review of advanced techniques for detecting plant diseases. Computers and Electronics in Agriculture 72, (1), pp. 1-13.

SAXTON, V., BOTTERILL, T., GREEN, R., AURAND, J.-M. (2014): First steps in translating human cognitive processes of cane pruning grapevines into AI rules for automated robotic pruning. BIO Web of Conferences 3, 01016.

SCHUELLER, J., WANG, M.-W. (1994): Spatially-variable fertilizer and pesticide application with GPS and DGPS. Computers and Electronics in Agriculture 11, (1), pp. 69-83.

SCHWARZ, H.-P., HEGE, D. (2014): GPS-supported weed control in open-field vegetable production. Agricultural technology 69, (2), pp. 68-71.

SEELAN, S., LAGUETTE, S., CASADY, G., SEIELSTAD, G. (2003): Remote sensing applications for precision agriculture: A learning community approach. Remote Sensing of Environment

88, (1-2), pp. 157-169.

SEITZ, M., ANGERMANN, D., BLOßFELD, M., DREWES, H., GERSTL, M. (2012): The 2008 DGFI realization of the ITRS: DTRF2008. Journal of Geodesy 86, (12), pp. 1097-1123.

STAFFORD, J. (2000): Implementing Precision Agriculture in the 21st Century. Journal of Agricultural Engineering Research 76, (3), pp. 267-275.

STAFFORD, J., AMBLER, B. (1994): In-field location using GPs for spatially variable field operations. Computers and Electronics in Agriculture 11, (1), pp. 23-36.

STEINIGER, S., HAY, G. (2009): Free and open source geographic information tools for landscape ecology. Ecological Informatics 4, (4), pp. 183-195.

STEINIGER, S., HUNTER, A. (2012): Free and Open source GIs software for Building a spatial Data Infrastructure. In: BOCHER, E., NETELER, M., eds, Geospatial Free and Open source software in the 21st Century, springer Berlin Heidelberg, pp. 247-261.

STEINIGER, S., HUNTER, A. (2013): The 2012 free and open source GIs software map - A guide to facilitate research, development, and adoption. Computers, Environment and Urban systems 39, pp. 136-150.

STEINIGER, S., WEIBEL, R. (2009): GIs software-a description in 1000 words. Encyclopeadia of Geography, 1-2.

SUN, H., SLAUGHTER, D., RUIZ, M., GLIEVER, C., UPADHYAYA, S., SMITH, R. (2010): RTK GPs mapping of transplanted row crops. Computers and Electronics in Agriculture 71, (1), pp. 32-37.

TANIGAKI, K., FUJIURA, T., AKASE, A., IMAGAWA, J. (2008): Cherry-harvesting robot. Computers and Electronics in Agriculture 63, (1), pp. 65-72.

TILLETT, N. (1991): Automatic guidance sensors for agricultural field machines:A review. Journal of Agricultural Engineering Research 50, pp. 167-187.

TOZER, P. (2009): Uncertainty and investment in precision agriculture - Is it worth the money? Agricultural systems 100, (1-3), pp. 80-87.

TRIMBLE NAVIGATION LIMITED (1998): 4700 Receiver - Operation Manual, sunnyvale.

TRIMBLE NAVIGATION LIMITED (2001): Trimble survey Controller - Reference Manual Volume 1, sunnyvale.

VANNUCCI, D., CINI, E., CIONI, A., VIERI, M. (1990): A prototype for the fully mechanized winter pruning of grape vines. Journal of Agricultural Engineering Research 46, pp. 1-11.

VOGELE, T. (2010): Mobile robots in agriculture - prospects and limits. KTBL, ed.

VOSS, K. (2012): What is actually... Remote sensing? - For advanced users.

WINDEGGER (2012): Neo Alpin - Operation and maintenance manual. WINDEGGER MASCHINEN S.R.L., ED.

WINDEGGER (2015): Neo Alpin - Product brochure.

WUNDER, E., KIELHORN, A., KLOSE, R., THIEL, M., RUCKELSHAUSEN, A. (2012): GIS- and sensor-based technologies for individual plant agriculture. Agricultural technology 67, (1), pp. 37-41.

ZARAGOZÍ, B., BELDA, A., LINARES, J., MARTINEZ-PEREZ, J., NAVARRO, J., ESPARZA, J. (2012): A free and open source programming library for landscape metrics calculations. Environmental Modeling & Software 31, pp. 131-140.

ZHANG, N., WANG, M., WANG, N. (2002): Precision agriculture-a worldwide overview. Computers and Electronics in Agriculture 36, (2-3), pp. 113-132.

ZHENG, L., ZHANG, J., WANG, Q. (2009): Mean-shift-based color segmentation of images containing green vegetation. Computers and Electronics in Agriculture 65, (1), pp. 93-98.

Internet sources

ADV (1995): Arbeitsgemeinschaft der Vermessungsverwaltungen - AdV-Online - Universale-Transversale-Mercator-Projektion, http://www.adv- online.de/Geodaetische-Grundlagen/UTM-Abbildung/. (02.01.2015).

ADV (2014): Arbeitsgemeinschaft der Vermessungsverwaltungen - AdV-Online - Universale-Transversale-Mercator-Projektion, http://www.adv- online.de/Geodaetische-Grundlagen/UTM-Abbildung/. (02.01.2015).

ADV (2015): Arbeitsgemeinschaft der Vermessungsverwaltungen - AdV-Online - Die GaiiLi-Kriiger-Abl | ıl(lung. http://www.adv-online.de/Geodaetische- Grundlagen/GK-Abbildung/. (02.01.2015).

AMAZONEN-WERKE (29.09.2011): BoniRob - Robots, http://go.amazone.de/?lang=1&news=26. (18.02.2015).

BKG (2014a): Geoportal.de - Geodata from Germany - Geoportal. FEDERAL AGENCY for CARTOGRAPHY AND GEODESY, ED., http://www.geoportal.de/DE/Geoportal/geoportal.html?lang=de. (21.12.2014).

BKG (2014b): Relationships between ellipsoidal height, normal height and quasigeoid. FEDERAL AGENCY FOR CARTOGRAPHY AND GEODESY, ED., http://gibs.bkg.bund.de/geoid/giderlaeuter.php. (31.12.2014).

BORK, S. (2008): Current status of free and open source desktop GIS, http://www2.htw-dresden.de/~fegis/DA/DA_BORK_2008/Vergleich.html. (21.12.2014).

BSO-GMBH (2014): General catalog 2014.

ESA (2013): The Earth's gravity field (geoid) as it will be seen by GOCE. EUROPEAN SPACE AGENCY, ED., http://www.esa.int/spaceinimages/Images/2004/10/The_Earth_s_gravity_field_geoid_as_it_will_be_seen_by_GOCE. (30.12.2014).

GDI SERVICE (2014): kvwmap Wiki, http://www.kvwmap.de/index.php/kvwmap_-_Open_Source_WebGIS. (21.12.2014).

GEOSN (2014): Basics and terms. STAATSBETRIEB GEOBASISINFORMATION UND VERMESSUNG SACHSEN, ED., http://www.landesvermessung.sachsen.de/inhalt/etrs/grund/grund.html. (02.01.2015).

OSNABRÜCK UNIVERSITY OF APPLIED SCIENCES (2014): GIS software. FACULTY OF AGRICULTURAL SCIENCES AND LANDSCAPE ARCHITECTURE, ED., https://www.al.hs-osnabrueck.de/11900.html. (21.12.2014).

HONDA (2015): Honda Engines | GX200 4-Stroke Engine | Features, Specs, and Model Info, http://engines.honda.com/models/model-detail/gx200. (27.02.2015).

MARTINET, A., BLAAUW, S. (2014): Field Robot Event Website, http://www.fieldrobot.nl/. (12.02.2015).

NcO (2015a): GPS.gov: control Segment. NATIONAL COORDINATION OFFICE FOR SPACE-BASED POSITIONING, NAVIGATION, AND TIMING, ED., http://www.gps.gov/systems/gps/control/. (05.01.2015).

NcO (2015b): GPS.gov: Selective Availability. NATIONAL COORDINATION OFFICE FOR SPACE-BASED POSITIONING, NAVIGATION, AND TIMING, ED., http://www.gps.gov/systems/gps/modernization/sa/. (04.01.2015).

NcO (2015c): GPS.gov: Space Segment. NATIONAL COORDINATION OFFICE FOR SPACE-BASED POSITIONING, NAVIGATION, AND TIMING, ED., http://www.gps.gov/systems/gps/space/. (04.01.2015).

NEUMAIR, S.-M., HAAS, H.-D.: Fernerkundung, http://wirtschaftslexikon.gabler.de/Archiv/9652/fernerkundung-v7.html. (18.02.2015).

NGA (2014): NGA GPS Ephemeris/Station/Antenna Offset Documentation. NATIONAL GEOSPATIAL-INTELLIGENCE AGENCY, ED., http://earth-info.nga.mil/GandG/sathtml/gpsdoc2014_12a.html. (04.01.2015).

STALLMAN, R. (2014): gnu.org, https://www.gnu.org/philosophy/free-sw.html. (21.12.2014).

UNIVERSITY OF ROSTOCK (2012): Geoinformatics Service, http://www.geoinformatik.uni-rostock.de/produkte_systemtyp.asp?ID_System=665081138. (21.12.2014).

VINE ROBOT (2015): VineRobot " Project, http://www.vinerobot.eu/project/. (12.02.2015).

VISION ROBOTICS (2015): Agriculture - Agriculture, http://www.visionrobotics.com/vrc/index.php?option=com_phocagallery&view=category&id=2:agriculture&Itemid=139. (12.02.2015).

8 Acknowledgments

With the conclusion of this work, I would like to thank all those who have made an extraordinary contribution to its success.

My special thanks go to Prof. Dr. habil. Hans-Peter Schwarz, head of the Institute of Technology at Hochschule Geisenheim University, for allowing me to work on the project. I was able to delve deeply into the very interesting and important future topic of robotics, especially in viticulture. I would also like to thank Prof. Dr. Schwarz for the great professional support and personal supervision during my affiliation with the institute and the processing time.

I would also like to thank Prof. Dr. habil. Bernd Honermeier, Head of the Institute of Crop Production and Plant Breeding I at Justus Liebig University in Gießen, for reviewing my dissertation.

I would like to thank the entire team at the Institute of Technology for their unique support over the last three years, especially the...

... Mrs. Gisela Berger as the good soul of the Institute for the organization and handling of all formalities and her warm, human personality.

. Dr. Rainer Keicher, as project coordinator, for his professional and personal support and his always open ear.

. Mr. Reinhold Müller for his support with the metrological issues and the preparation of the technology used.

... Mr. Marcel Scherber for his technical and personal support during the design work and tests.

... Mr. Hans Stortlander, Mr. Gunther Schrauth and Mr. Sandor Hum for their help and excellent cooperation.

. and especially Mr. David Brunner, without whom the control of the PHENObot would not have been possible in this form. Mr. Brunner played a very large part in the success of the project. With his selfless nature and his open ear, he was always there to help me with many problems, even late at night. Many thanks David!

I would also like to thank my colleagues at the Julius Kühn Institute for Grapevine Breeding Geilweilerhof, in this case Dr. Anna Kicherer, for their constructive project work and successful management of the trials. I would also like to thank all the other partners in the PHENOvines research project, the Heinrich Mayer company, the German Winegrowers' Cooperative and the Reichhardt company. The same applies to the Federal Ministry of Education and Research and the project sponsor Jülich for funding the project.

I would also like to thank Mr. Andreas Kofler, KMS Mechatronics, and Mr. Stefan Schiefer, Windegger, for their technical and human support.

My biggest thanks go to my parents, my family, my friends, but above all to my wife Susanne and our daughters Emma and Hannah, who have always supported and motivated me during my training and beyond. Thank you for your patience!

Eimsheim

Philipp Rüger

9 Appendix

Due to the large volume, the individual contents are not printed. Instead, all additional documents, measurement data, evaluations, etc. are available as a digital appendix in the form of a ZIP archive on the server of the Gießen Electronic Library (GEB):

http://geb.uni-giessen.de/geb

Kolibri 360 GmbH
Ferdinand-Jühlke-Straße 7
99095 Erfurt, Deutschland
produktsicherheit@kolibri360.de

Druck:
CPI Druckdienstleistungen GmbH
im Auftrag der
Kolibri 360 GmbH
Ferdinand-Jühlke-Str. 7
99095 Erfurt